FOCUS on DECIMALS

Revised

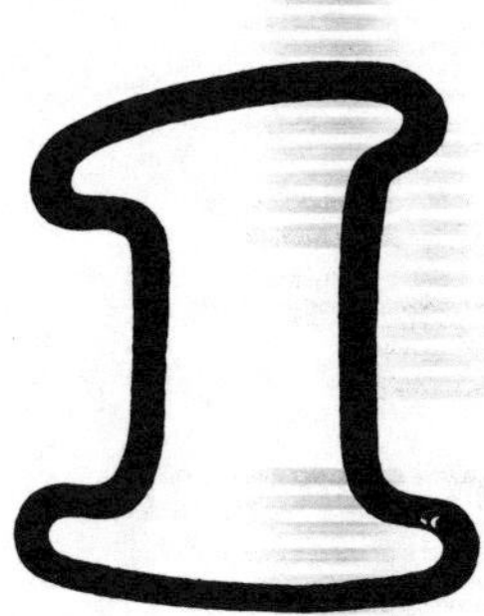

63.48 51

By Margaret A. Smart

FOREWORD:

Focus on Decimals, Book 1 is a carefully sequenced book of student activities for mastering the meaning of decimals and how to compare, add and subtract them. Decimals are presented as an extension of the number system. Congruent regions introduce tenths, then hundredths, and finally thousandths. The same types of diagrams enables students to compare decimals, and make the generalizations or rules meaningful. Many activities are included in the form of puzzles and patterns. Pictures are used to develop the two operations of addition and subtraction. Once the technique for addition and subtraction are established, money is presented to give meaning to operating with hundredths.

There is a companion volume on multiplication and division.

Mary Laycock
Mathematics Specialist
Nueva Center for Learning
Hillsborough, California 94010

INTRODUCTION:

Focus on Decimals is a set of two booklets designed to give the student a successful experience in decimals. The set is geared toward increasing the computational ability of the student.

The set can be used in individualized programs, or in the more traditional type program. The booklets were written so that the student can work on his own with a minimum of teacher assistance. Quizzes, puzzles, and review pages are included. It is suggested that manipulative materials and lead-up games be used along with the program.

These materials were designed for use in grades five to nine.

Margaret A. Smart

P.O. Box 4875
Hayward, California 94540

FOCUS ON DECIMALS

BOOK 1

ADDITION AND SUBTRACTION

by

Margaret A. Smart

CONTENTS

THE PLACE VALUE CHART

Have you seen a place value chart for whole numbers? It looks like this.

millions	hundred thousands	ten thousands	thousands	hundreds	tens	ones
7	0	2	6	0	3	5

Each digit on the chart has a special place value name.

The **5** means 5 ones.

The **7** means 7 millions.

We will expand the place value chart to include decimals. This is how the chart looks with whole numbers and decimals.

WHOLE NUMBERS								DECIMALS					
millions	hundred thousands	ten thousands	thousands	hundreds	tens	ones		tenths	hundredths	thousandths	ten thousandths	hundred thousandths	millionths
7	0	2	6	0	3	5	.	8	4	9	1	0	6

What is the place name for the digit 8? ____________________

What is the place name for the digit 4? ____________________

What is the place name for the digit 9? ____________________

Did you write tenths, hundredths, and thousandths? ____________________

We will learn about all kinds of decimal numbers in this book. We will start with tenths.

TENTHS

How many parts? __________

How many shaded parts? __________

3/10 is the fraction name.

0.3 is the decimal name for 3/10.

DECIMALS ARE ANOTHER WAY OF WRITING FRACTIONS. THE DOT WE USE IS CALLED THE DECIMAL POINT.

Here are some examples.

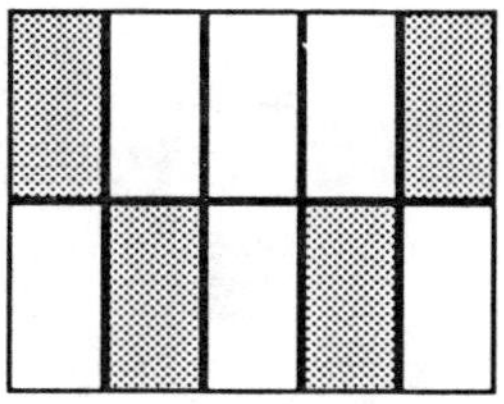

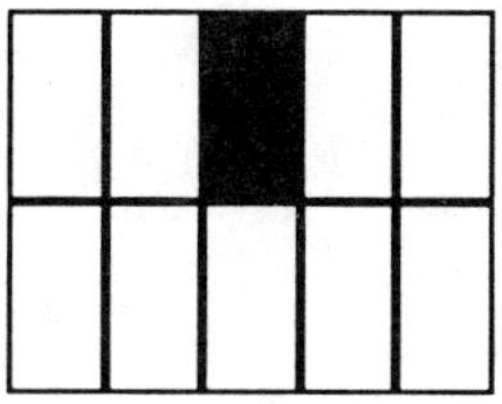

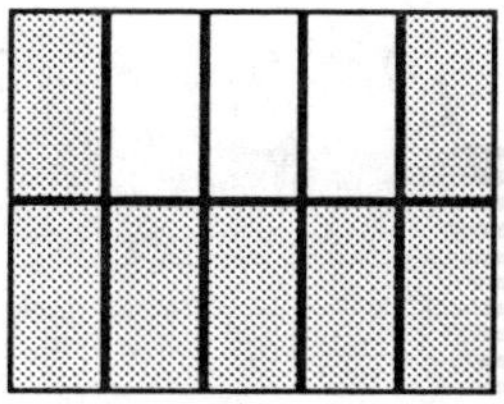

Fraction name:	4/10	1/10	7/10
Decimal name:	0.4	0.1	0.7

Try these:

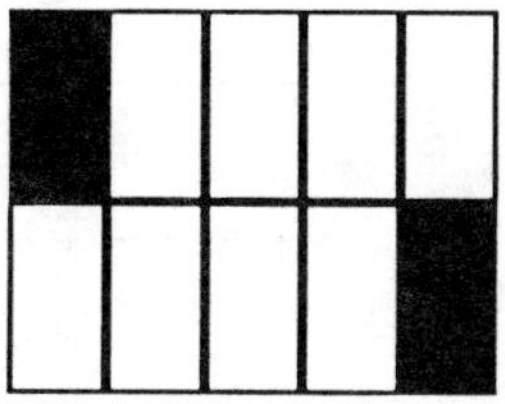

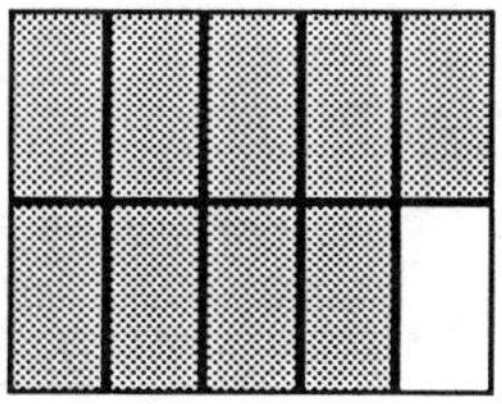

Fraction name:	______	______	______
Decimal Name:	______	______	______

A RULE FOR TENTHS

TENTHS HAVE ONLY ONE ZERO IN THE DENOMINATOR. THERE IS ONE DIGIT AFTER THE DECIMAL POINT.

3 /10	**=**	**0.03**
1 zero		**1 digit**

Write these fractions as decimals.

4/10	=	0.4	5/10	=	_____	8/10	=	_____
7/10	=	_____	9/10	=	_____	2/10	=	_____

Write these decimals as fractions.

0.2	=	2/10	0.7	=	_____	0.1	=	_____
0.3	=	_____	0.6	=	_____	0.9	=	_____

Complete this chart.

DECIMAL	FRACTION	WE READ AND SAY
0.1	1/10	one tenth
0.3		
0.5		
	4/10	
		two tenths
	7/10	
		nine tenths

WRITING WHOLE NUMBERS AS DECIMALS

ONE WHOLE

How many shaded parts? _____

10/10 is the fraction name.

Does 10/10 name one whole? _____

> **1.0 is the decimal name for one whole.**
> **We place a zero after the decimal point to show that there are no tenths.**

Write these whole numbers as decimal numbers.

124	124.0	44	_____	98	_____
621	_____	67	_____	11	_____
137	_____	25	_____	39	_____

Write these decimal numbers as whole numbers.

36.0	36	135.0	_____	21.0	_____
19.0	_____	28.0	_____	9.0	_____
100.0	_____	5.0	_____	82.0	_____

> **NUMBERS SUCH AS 34 AND 34.0 HAVE THE SAME MEANING.**
> **IN OUR LATER WORK IT WILL BE USEFUL TO KNOW HOW TO WRITE A WHOLE NUMBER AS A DECIMAL NUMBER.**

MIXED NUMBERS

THE DECIMAL POINT IS USED TO SEPARATE WHOLE NUMBERS AND FRACTIONS.

This is a picture of the mixed number one and three tenths.

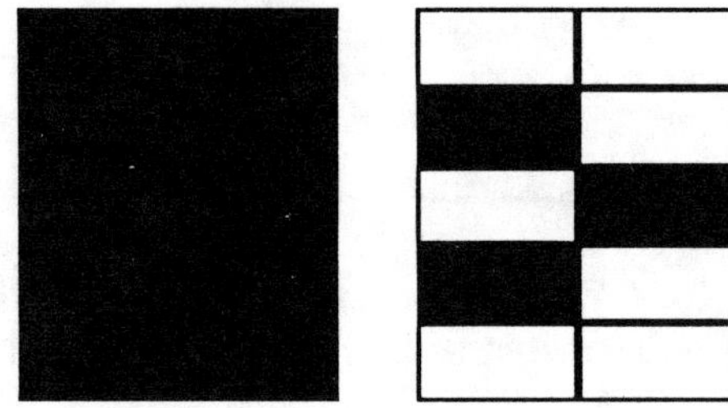

1 3/10 is the fraction name.

1. 3 is the decimal name.

Write mixed numbers for each set of pictures.

Fraction name: ______

Decimal name: ______

Fraction name: ______

Decimal name: ______

Fraction name: ______

Decimal name: ______

Fraction name: ______

Decimal name: ______

PRACTICE PAGE

Write these fractions as decimals.

5/10	_____	9/10	_____	8/10	_____
7/10	_____	4/10	_____	1/10	_____
3/10	_____	6/10	_____	2/10	_____

Did you remember to put the zero before the decimal point?

Write these whole and mixed numbers as decimals.

17 2/10	= _____	18 7/10	= _____	33 8/10	= _____
11	= _____	29 4/10	= _____	128 1/10	= _____
25 3/10	= _____	2 5/10	= _____	93	= _____

Write these decimals as fractions.

2.1	= _____	19.6	= _____	0.8	= _____
15.1	= _____	0.7	= _____	123.6	= _____
17.4	= _____	0.3	= _____	12.8	= _____

Complete the following sentences.

The dot we use in decimals is called the decimal _______________.

The decimal _______________ is used to separate whole numbers and _______________.

A zero placed after the decimal point shows that there are no _______________.

PUZZLE PAGE

Write a decimal for the shaded part of each picture.

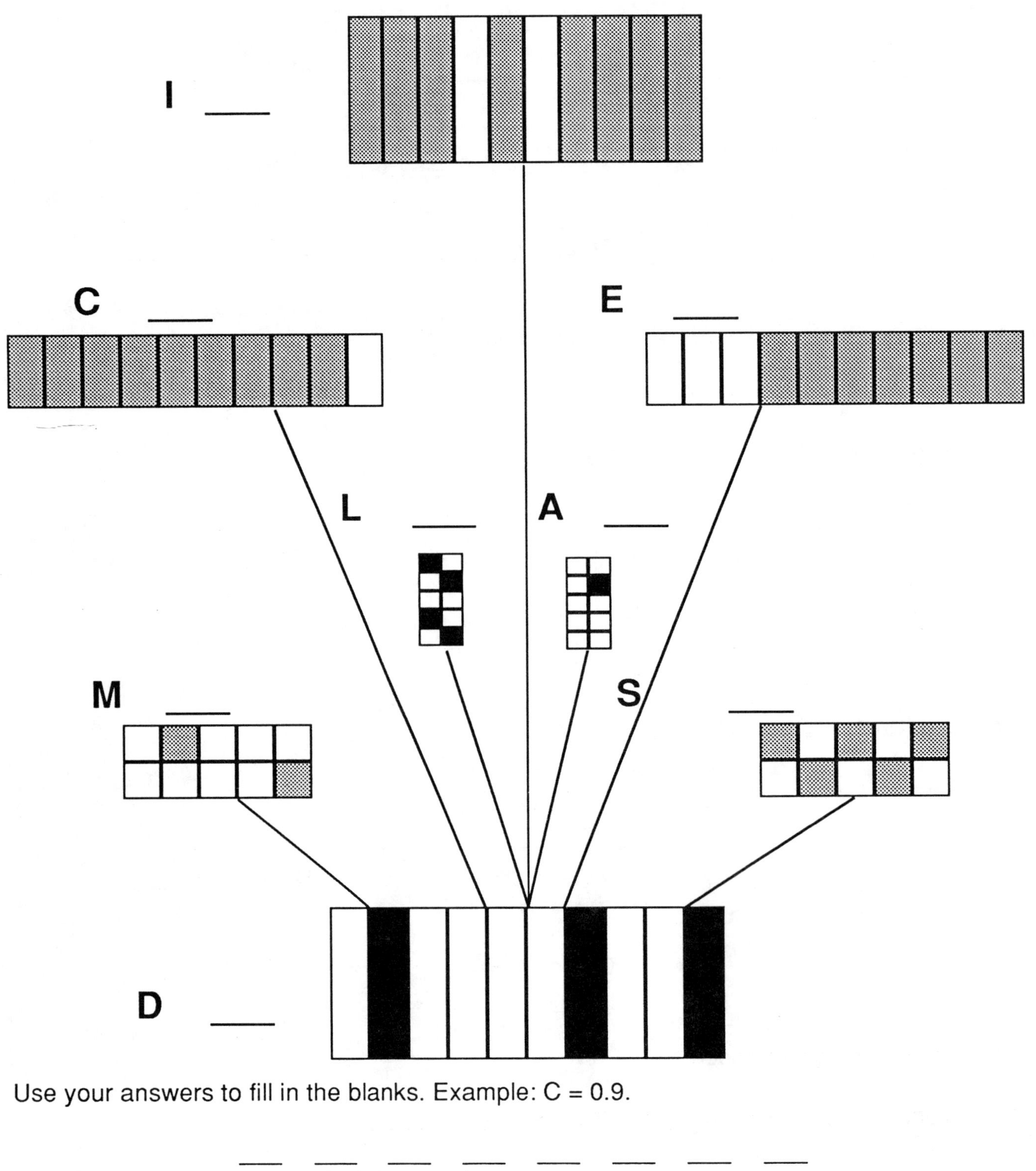

Use your answers to fill in the blanks. Example: C = 0.9.

___	___	___	___	___	___	___	___
0.3	0.7	0.9	0.8	0.2	0.1	0.4	0.5

DECIMALS AND PLACE VALUE

WHOLE NUMBERS	DECIMAL NUMBERS	MIXED NUMBERS
3, 14, 280, 78	0.3, 0.7, 0.9, 0.5	100.3, 1.2, 14.8, 2.4

All of the numbers in the boxes have been written in the place value chart.
On the chart below:

Put an "X" by the whole numbers.
Put a circle by the decimal numbers.
Put a check by the mixed numbers.

	millions	hundred thousands	ten thousands	thousands	hundreds	tens	ones		tenths	hundredths	thousandths	ten thousandths	hundred thousandths	millionths
X							3	.						
O							0	.						
√							1	.	2					
						1	4	.	8					
							0	.	7					
						1	4	.						
							0	.						
					2	8	0	.						
					1	0	0	.	3					
							0	.	5					
							2	.	4					
						7	8	.						

DECIMALS AND PLACE VALUE

Part of the place value chart is below. Match each item below with a number that is written on the place value chart. Use a ruler to draw lines like those in the example.

PLACE VALUE CHART

- forty- three
- six-tenths
- four and three-tenths
- six
- eight and three-tenths
- nine tenths
- eighty-three
- nine
- two hundred twenty one
- nine and two tenths
- three tenths
- nineteen and nine tenths
- seven tenths
- seven and four tenths
- one tenth
- one

thousands	hundreds	tens	ones		tenths	hundredths	thousandths
			9	.			
		8	3	.			
			0	.	9		
			8	.	3		
			6	.	0		
			4	.	3		
			0	.	6		
		4	3	.			
			1	.			
			0	.	1		
			7	.	4		
			0	.	7		
		1	9	.	9		
			0	.	3		
			9	.	2		
	2	2	1	.			

Take Quiz 1.

HUNDREDTHS

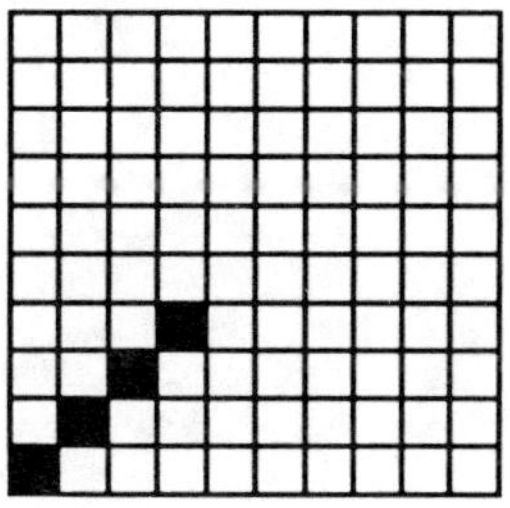

There are **100** parts in this whole.

How many are shaded? ________

4/100 is the fraction name.

0.04 is the decimal name for 4/100.

Fraction name:	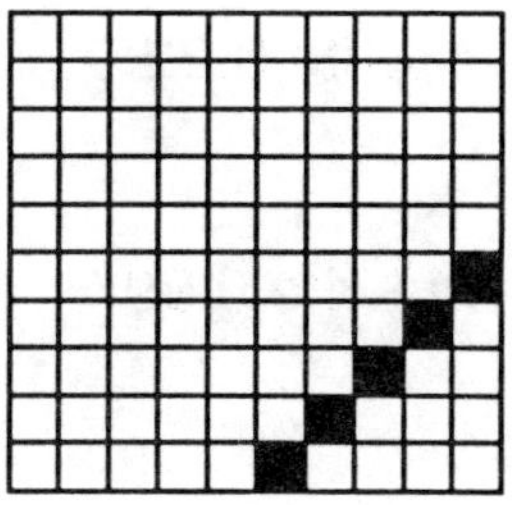 5/100	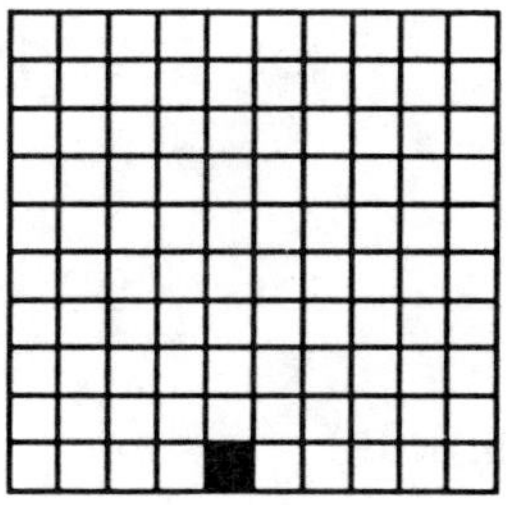 1/100	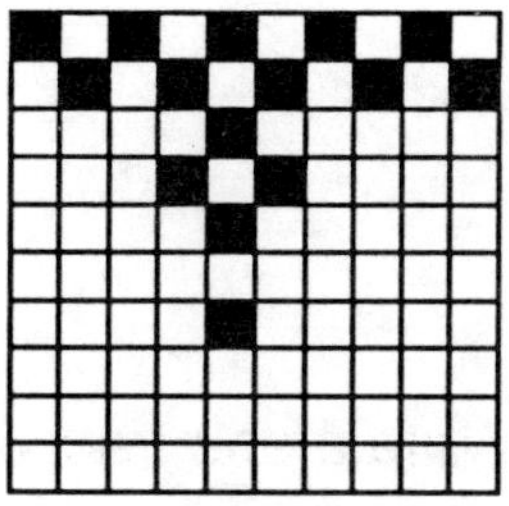 15/100
Decimal name:	0.05	0.01	0.15

Try these:

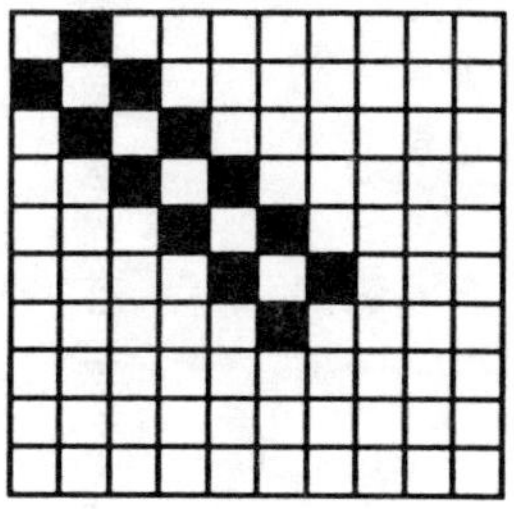

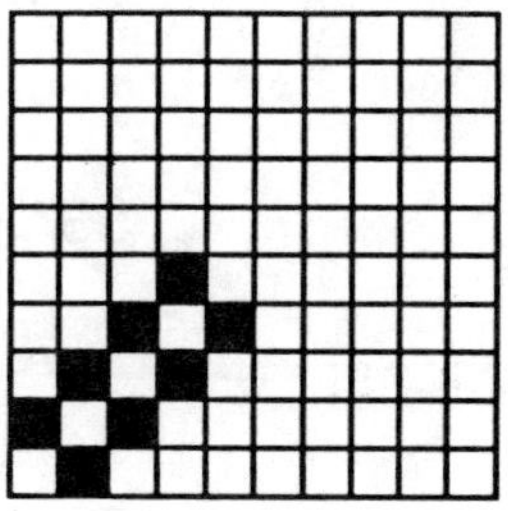

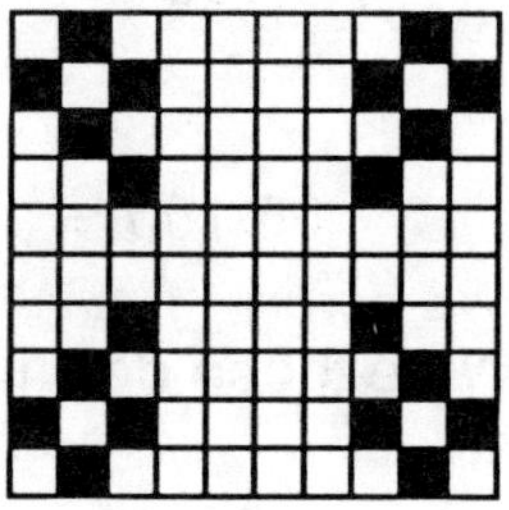

Fraction name:	________	________	________
Decimal name:	________	________	________

A RULE FOR HUNDREDTHS

HUNDREDTHS HAVE TWO ZEROS IN THE DENOMINATOR. THERE ARE TWO DIGITS AFTER THE DECIMAL POINT.

8/100 = 0.08

2 zeros 2 digits

This is how the rule works:

for 1/100 we write 0.01

for 22/100 we write 0.22

for 99/100 we write 0.99

Write these fractions as decimals:

4/100 = _____	8/100 = _____	45/100 = _____
19/100 = _____	34/100 = _____	15/100 = _____
3/100 = _____	7/100 = _____	11/100 = _____
66/100 = _____	53/100 = _____	91/100 = _____

Write these decimals as fractions:

0.04 = _____	0.75 = _____	0.82 = _____
0.02 = _____	0.55 = _____	0.64 = _____
0.12 = _____	0.09 = _____	0.13 = _____

MORE WORK WITH HUNDREDTHS

Complete this chart:

DECIMAL	FRACTION	WE READ AND SAY
0.02	2/100	two hundredths
0.11	11/100	eleven hundredths
0.24	________	____________________
________	________	fifty hundredths
________	12/100	____________________
________	9/100	____________________
________	46/100	____________________
0.07	________	____________________

Each item in the box can be matched with an item below. Write your answer in the space.

65/100	0.14	12/100	0.75	53/100
sixteen hundredths		0.44	four hundredths	

4/100 ________________ 16/100 ________________ 44/100 ________________

0.65 ________________ 0.53 ________________ 0.12 ________________

fourteen hundredths ____________ seventy-five hundredths ____________

THOUSANDTHS

This is a picture of one whole.

It has 1000 equal parts.

How many parts are shaded?__________

7/1000 is the fraction name.

0.007 is the decimal name.

ONE WHOLE

This is a picture of one whole. It also has 1000 equal parts.
How many shaded parts? ________

The fraction name for shaded parts is ________

The decimal name for the shaded parts is 0.024.

How many unshaded parts? ________

The fraction name for the unshaded parts is _____

The decimal name for the unshaded parts is _____

A RULE FOR THOUSANDTHS

THOUSANDTHS HAVE THREE ZEROS IN THE DENOMINATOR. THERE ARE THREE DIGITS AFTER THE DECIMAL

7 /1000	**=**	**0.007**
3 ZEROS		**3 DIGITS**

This is how the rule works:

for 3/1000 we write 0.003

for 71/1000 we write 0.071

for 288/1000 we write 0.288

Write these fractions as decimals.

8/1000 = 0.008	16/1000 = _____	32/1000 = _____
4/1000 = _____	122/1000 = _____	65/1000 = _____
81/1000 = _____	5/1000 = _____	44/1000 = _____
2/1000 = _____	75/1000 = _____	343/1000 = _____

Write these decimals as fractions.

0.003 = _____	0.012 = _____	0.015 = _____
0.875 = _____	0.128 = _____	0.064 = _____
0.009 = _____	0.102 = _____	0.093 = _____
0.098 = _____	0.644 = _____	0.333 = _____

MORE WORK WITH THOUSANDTHS

Complete this chart.

DECIMAL	FRACTION	WE READ AND SAY
0.004	4/1000	four thousandths
0.023	23/1000	twenty-three thousandths
0.132	________	____________________________
________	________	ten thousandths
________	42/1000	____________________________
________	101/1000	____________________________
________	17/1000	____________________________
0.622	________	____________________________

REVIEW EXERCISES

Fill in the blanks.

0.13	names	13	hundredths
0.7	names	7	tenths
0.73	names	___	____________
0.791	names	___	____________
0.108	names	___	____________
0.63	names	___	____________
0.8	names	___	____________

PICTURE PUZZLE

53/100	1 5/1000	1 5/10	3/100
53/1000	1/100	7/10	4/100
1/10	1 5/100	1/1000	7/100

Match each number below with one number in the box above. Cut out one piece at a time and paste it in the correct box.

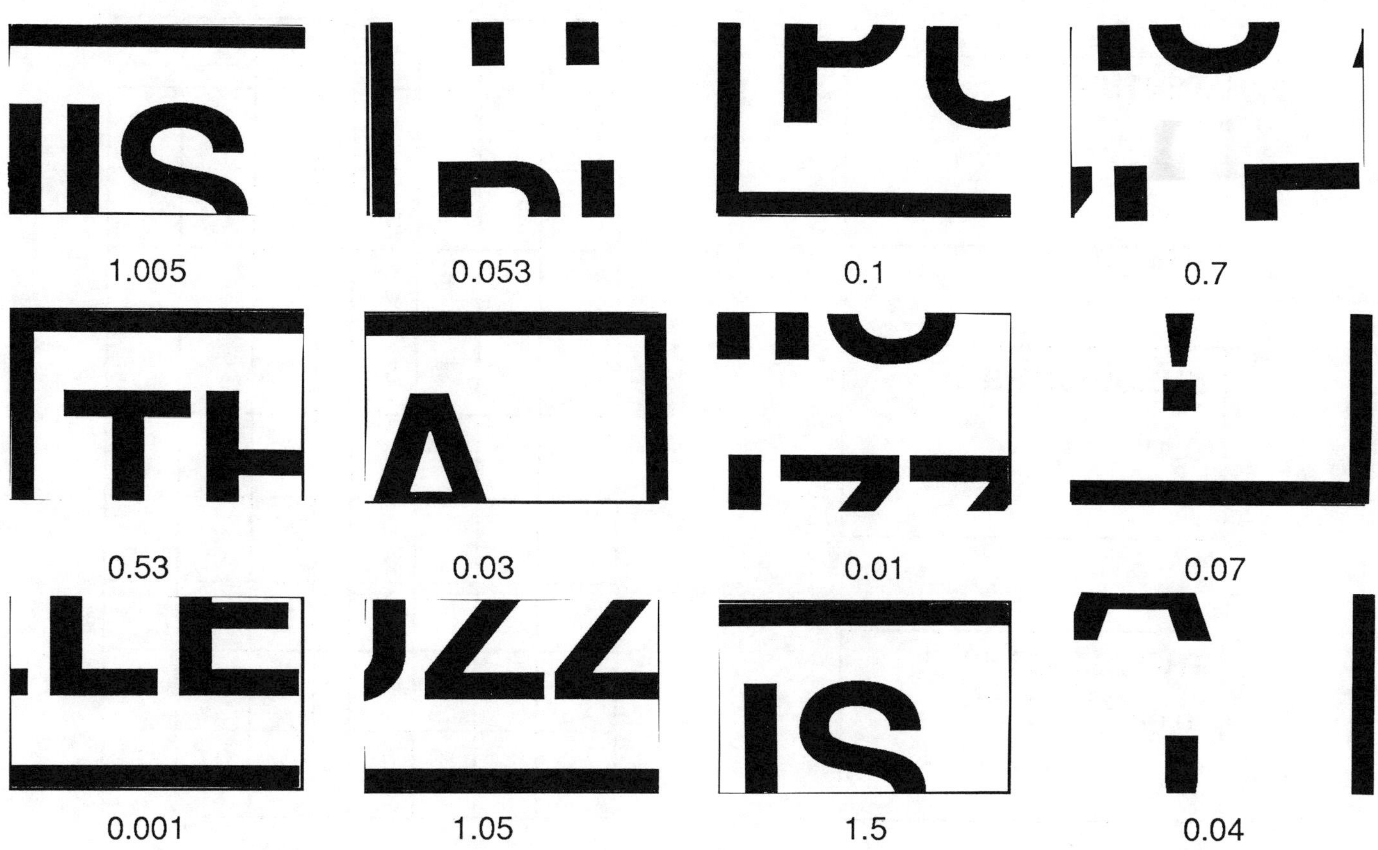

MORE WORK WITH PLACE VALUE

Part of the place value chart is shown below.

Write each number that is on the **Place Value Chart** in one of the boxes below. You should have three numbers in each box.

WHOLE NUMBERS

6

MIXED NUMBERS

TENTHS

HUNDREDTHS

0.41

THOUSANDTHS

0.085

PLACE VALUE CHART

thousands	hundreds	tens	ones		tenths	hundredths	thousandths
			6	.			
			0	.	0	8	5
			0	.	4	1	
	1	2	2	.	3		
			0	.	0	7	
			0	.	4		
		1	0	.			
			0	.	1	1	8
			0	.	9		
	1	7	8	.			
		1	0	.	4	1	
			0	.	6	7	
			0	.	1		
			4	.	2	1	7
			0	.	0	0	7

MORE WORK WITH PLACE VALUE

Match each item below with a number that is written on the Place Value Chart. Use a ruler to draw lines between matching points.

- two hundred two thousandths
- three thousandths
- two and four tenths
- five hundredths
- ten and three thousandths
- twenty-two
- two and forty hundredths
- fifty-one thousandths
- nineteen hundredths
- one hundred and one tenth
- nine tenths
- six hundred
- twenty-seven thousandths
- eighty-eight hundredths
- eight tenths
- eleven hundredths
- one tenth
- sixty

PLACE VALUE CHART

thousands	hundreds	tens	ones		tenths	hundredths	thousandths
			0	.	0	0	3
			0	.	2	0	2
		2	2	.			
		1	0	.	0	0	3
			0	.	0	5	
			2	.	4		
			0	.	0	5	1
			2	.	4	0	
	6	0	0	.			
			0	.	9		
	1	0	0	.	1		
			0	.	1	9	
			0	.	8	8	
			0	.	0	2	7
		6	0	.			
			0	.	1		
			0	.	1	1	
			0	.	8		

COMPARING TENTHS

$\frac{1}{10}$	$\frac{2}{10}$	$\frac{3}{10}$	$\frac{4}{10}$	$\frac{5}{10}$	$\frac{6}{10}$	$\frac{7}{10}$	$\frac{8}{10}$	$\frac{9}{10}$	$\frac{10}{10}$
0.1	0.2	0.3	0.4	0.5	0.6	0.7	0.8	0.9	1.0

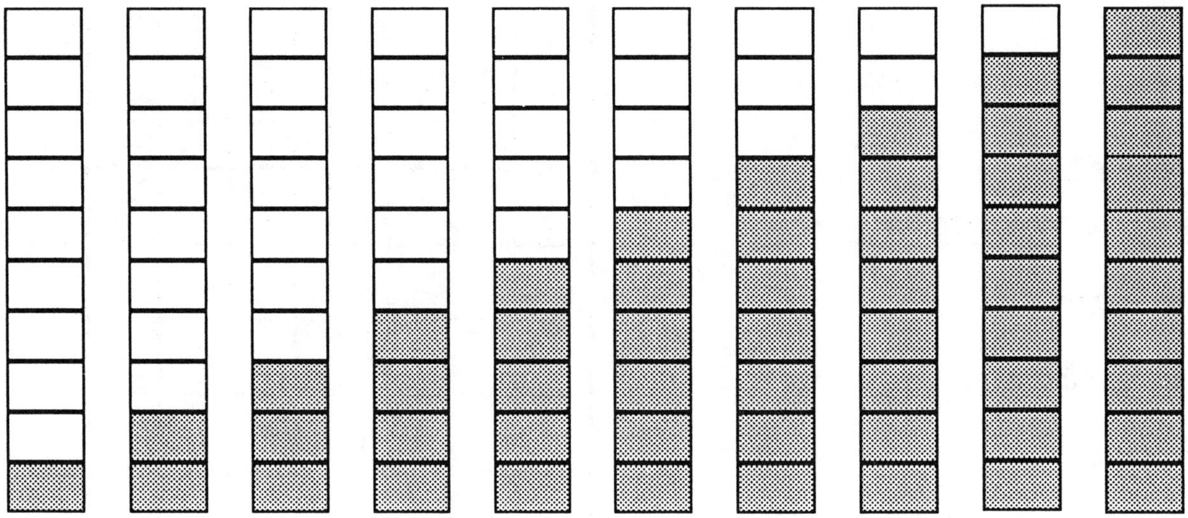

COMPARING SYMBOLS

> means larger or greater than

< means smaller or less than

= means equal to

Use the symbols in the chart to compare the first decimal with the second.

Example: **0.3 < 0.9**

0.2	________	0.4	0.5	________	0.1	0.7	________	0.9
0.8	________	0.5	0.8	________	0.7	0.3	________	0.4
0.6	________	1.0	1.0	________	0.9	0.9	________	0.6

COMPARING HUNDREDTHS

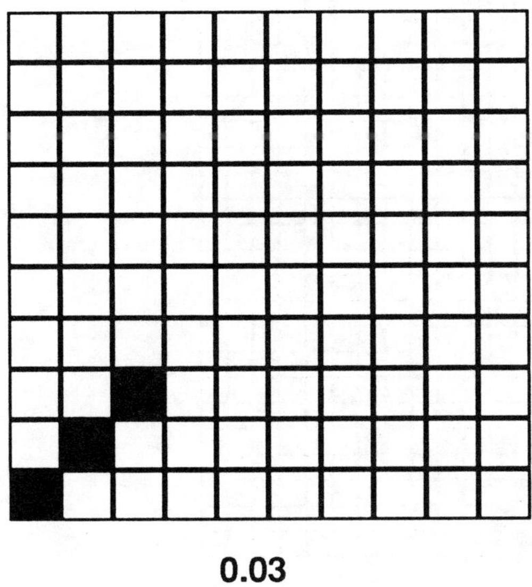

0.03

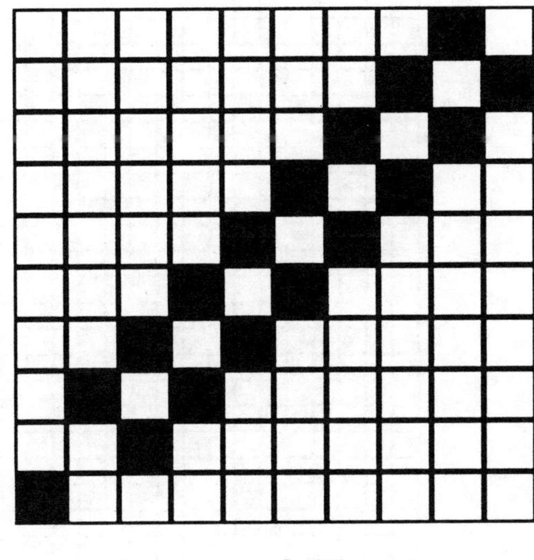

0.17

How many hundredths are shaded in the first picture? _______

How many hundredths are shaded in the second picture? _______

Compare the first picture with the second picture. Is 3 larger or smaller than 17? _____________ Then we know that three hundredths is smaller than seventeen hundredths.

WE WRITE: 0.03 < 0.17.

Compare the first decimal with the second decimal in the problems below. Use **<, > or =** in the blank.

0.61 ______ 0.72   0.81 ______ 0.29   0.65 ______ 0.03

0.05 ______ 0.99   0.47 ______ 0.46   0.10 ______ 0.11

0.01 ______ 0.10   0.76 ______ 0.67   0.26 ______ 0.09

COMPARING THOUSANDTHS

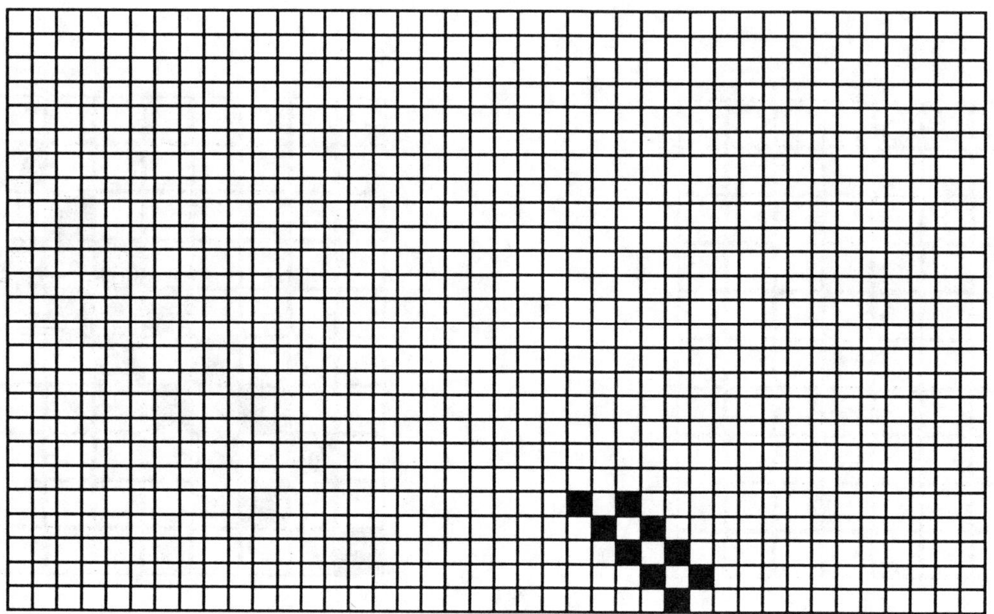

How many thousandths are shaded in the picture?

Write a decimal (in thousandths) for the shaded part.

How many thousandths are **not** shaded in the picture?

Write a decimal (in thousandths) for the unshaded part.

To compare the shaded parts with the unshaded parts, we write:
0.009 < 0.991

(Nine thousandths is less than nine hundred ninety-one thousandths.)

Compare the first decimal with the second decimal in the problems below. Use <, > or = in the blank.

0.123	>	0.011	0.542	____	0.427	0.198	____	0.246
0.777	<	0.982	0.105	____	0.021	0.100	____	0.101
0.823	____	0.864	0.287	____	0.159	0.453	____	0.509
0.260	____	0.765	0.333	____	0.007	0.559	____	0.995
0.011	____	0.010	0.423	____	0.772	0.999	____	0.990

COMPARING TENTHS AND HUNDREDTHS

The shaded part is the same in these two pictures.

TENTHS

HUNDREDTHS

Fraction name:	1/10	10/100
Decimal name:	0.01	0.10

Write a fraction and a decimal for the shaded parts.

TENTHS

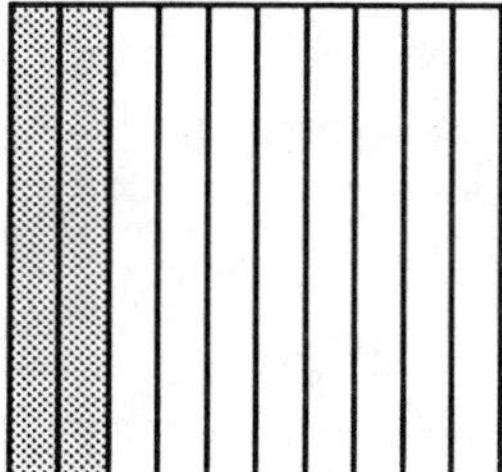

HUNDREDTHS

Fraction name:	__/10	___/100
Decimal name:	________	________

Write these tenths as hundredths:

0.4	=	________	0.9	=	________	0.5	=	________
0.7	=	________	0.3	=	________	0.6	=	________
0.2	=	________	0.8	=	________	0.1	=	________

TO CHANGE TENTHS TO HUNDREDTHS, WE ADD A ZERO AFTER THE TENTHS DIGIT.

0.3 = 0.30

COMPARING TENTHS AND HUNDREDTHS

0.3 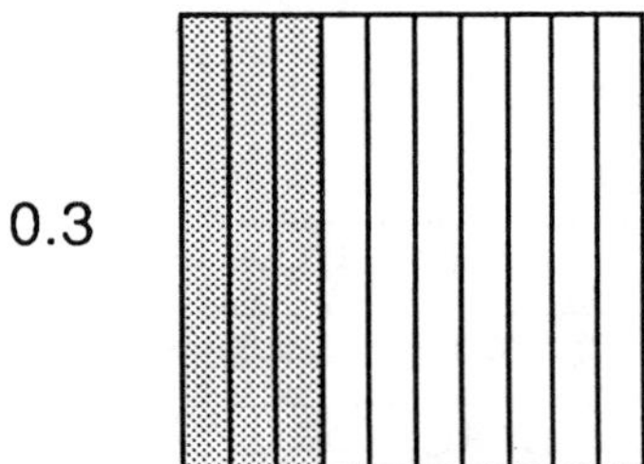 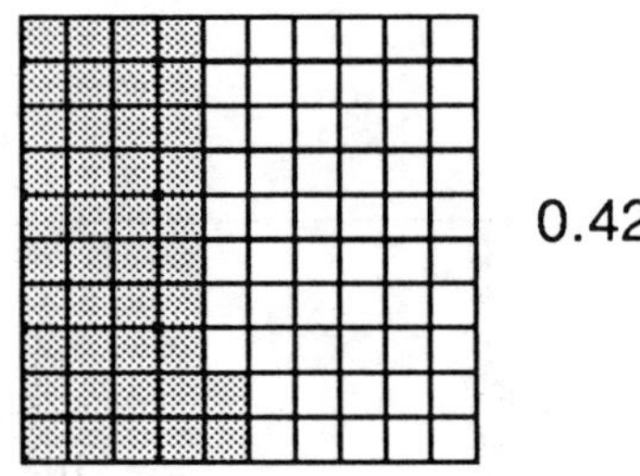 0.42

Is 0.3 larger or smaller than 0.42? __________

To compare these pictures we will change tenths to hundredths.

0.3 ________ 0.42

0.30 ________ 0.42

Remember: To change tenths to hundredths, we add a zero to the tenths digit.

Compare the first decimal with the second decimal. Change tenths to hundredths. Put **<, > or** = in the space.

0.5 ___?___ 0.19

0.50 ___>___ 0.19

0.26 ___?___ 0.7

____

0.89 ___?___ 0.9

____

0.67 ___?___ 0.7

____

0.3 ___?___ 0.37

____

0.9 ___?___ 0.92

____

0.12 ___?___ 0.1

____

0.6 ___?___ 0.60

____

COMPARING TENTHS, HUNDREDTHS, AND THOUSANDTHS

ADDING ZEROS TO DIGITS AFTER THE DECIMAL POINT DOES NOT CHANGE THE VALUE OF THE DECIMAL.
ALL OF THESE DECIMALS ARE EQUAL:-

0.2 = 0.20 = 0.200

These examples show you how zeros are used in comparing decimal numbers.

0.3	________	0.219	0.24	________	0.214	0.6	________	0.62
0.300	________	0.219	0.240	________	0.214	0.60	________	0.62

Compare the first decimal with the second decimal. Add zeros where needed. Use **<, > or** = in the blank.

0.31	________	0.762	0.86	________	0.727
0.33	________	0.241	0.2	________	0.19
0.106	________	0.11	0.4	________	0.234
0.91	________	0.929	0.541	________	0.5
0.22	________	0.222	0.218	________	0.3
0.63	________	0.631	0.99	________	0.9
0.1	________	0.101	0.3	________	0.299
0.4	________	0.400	0.94	________	0.940
0.234	________	0.23	0.105	________	0.11
0.22	________	0.219	0.876	________	0.87

Take Quiz 2.

PICTURING TENTHS IN ADDITION AND SUBTRACTION

We can use pictures to add and subtract decimals.

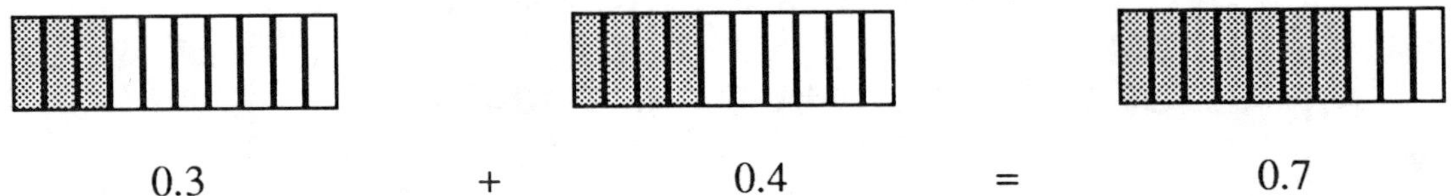

0.3 + 0.4 = 0.7

Write the decimal number for each sum and difference. Watch the signs.

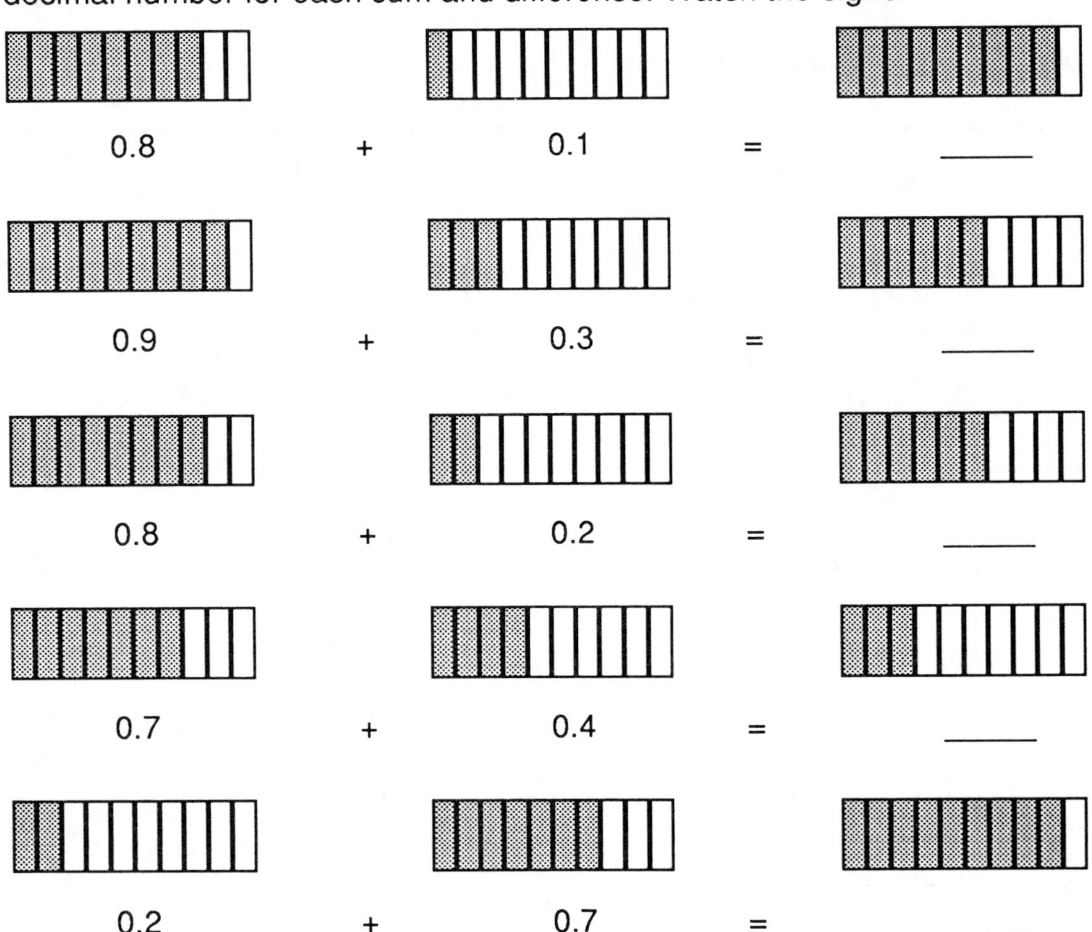

0.8 + 0.1 = _____

0.9 + 0.3 = _____

0.8 + 0.2 = _____

0.7 + 0.4 = _____

0.2 + 0.7 = _____

Shade in the correct number of tenths for these addition problems.

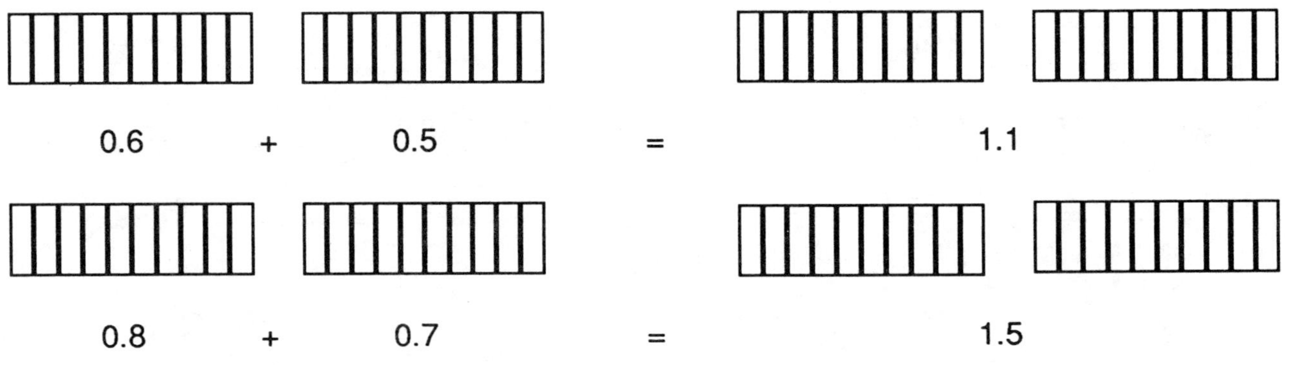

0.6 + 0.5 = 1.1

0.8 + 0.7 = 1.5

PICTURING TENTHS IN ADDITION AND SUBTRACTION

We can arrange pictures another way to add and subtract decimals.

+

0.4
+ 0.5
0.9

-

1.0
- 0.6
0.4

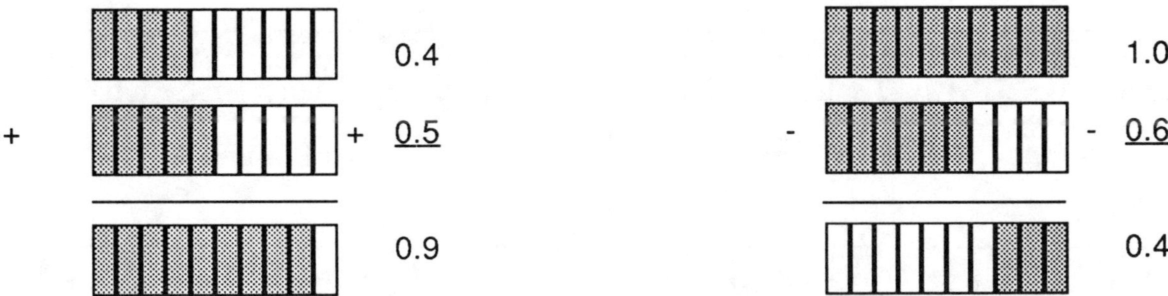

Write the decimal number for each sum and difference. Shade in the correct number of tenths for each answer.

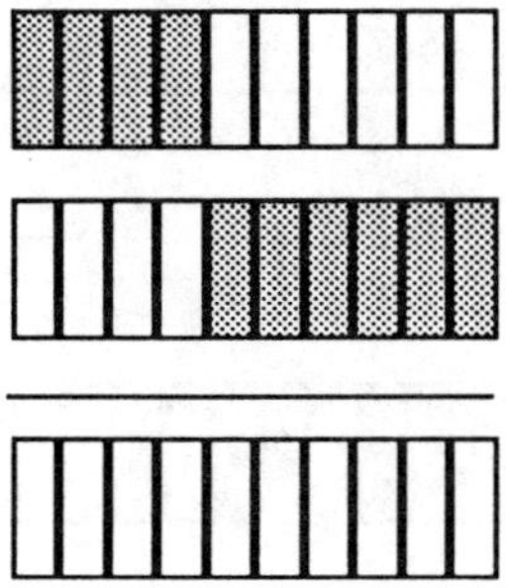

+

0.4
+ 0.6

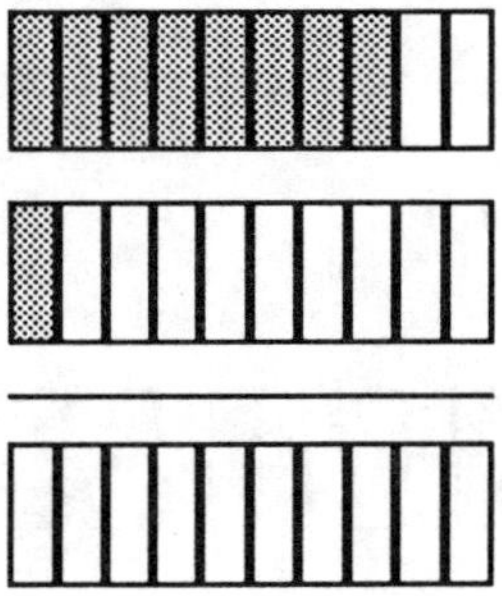

-

0.8
- 0.1

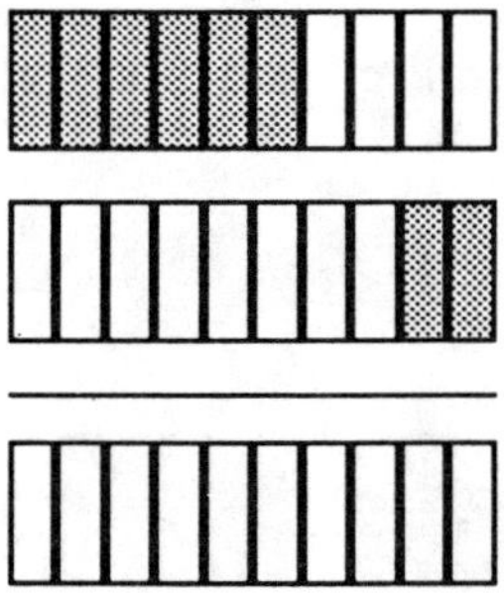

+

0.6
+ 0.2

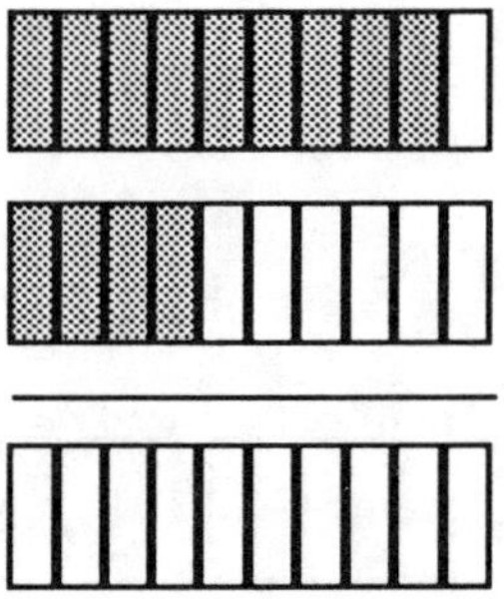

-

0.9
- 0.5

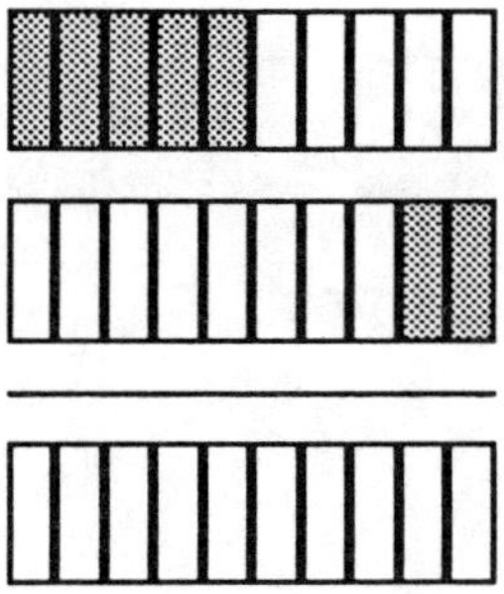

+

0.5
+ 0.2

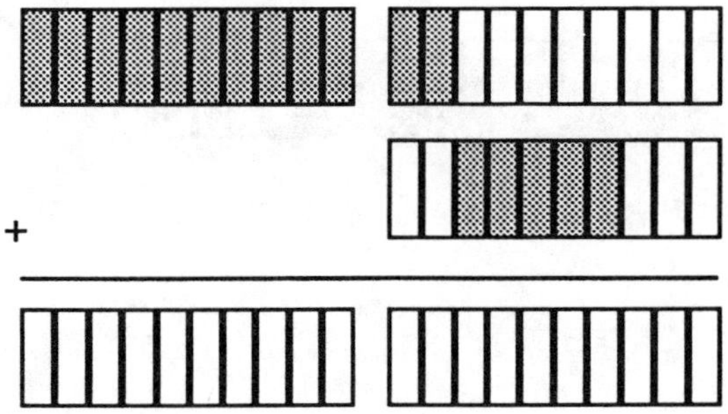

+

1.2
+ 0.5

PICTURING REGROUPING IN ADDITION

Write the number of tenths and the decimal for each sum. An example has been done for you.

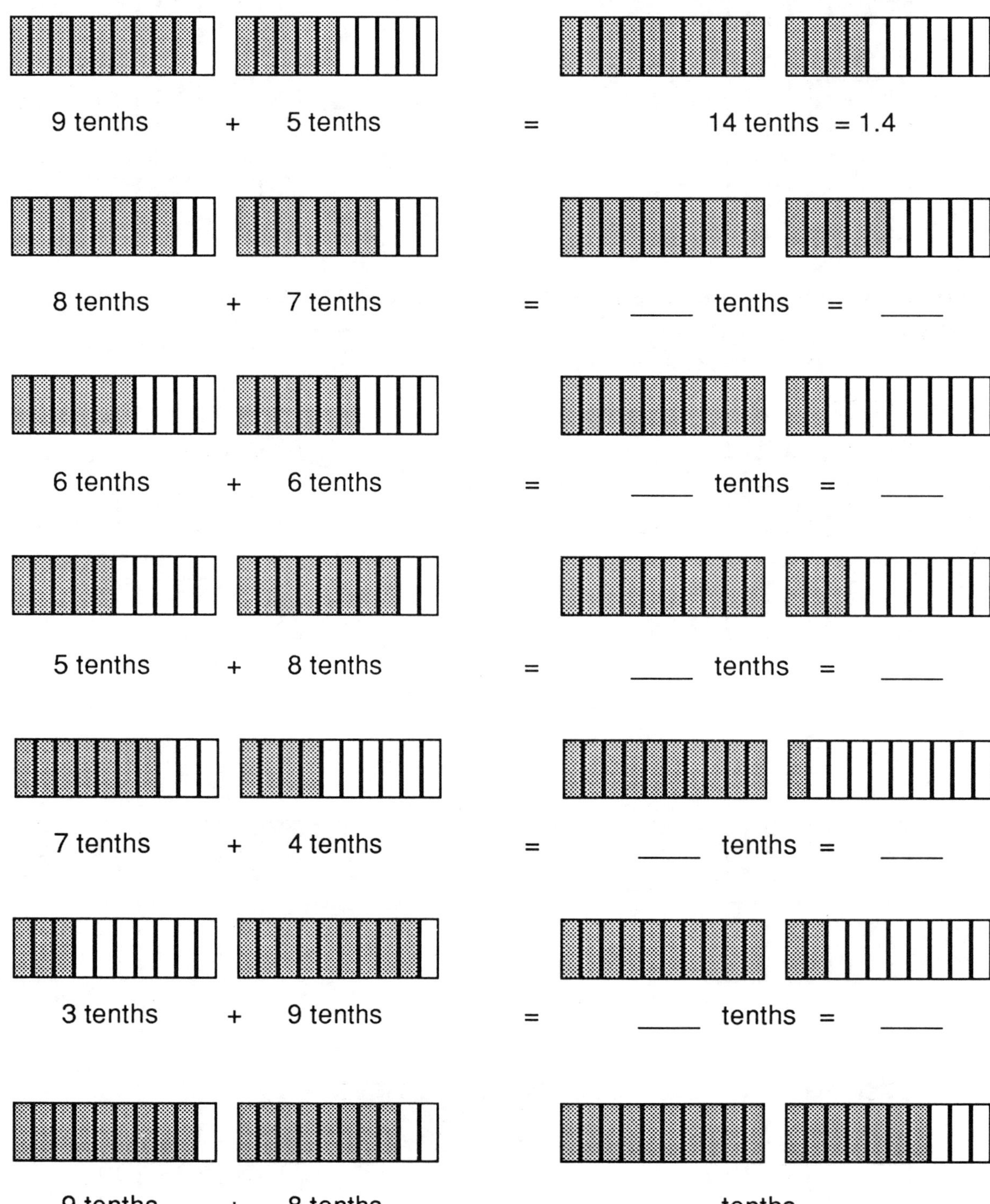

9 tenths + 5 tenths = 14 tenths = 1.4

8 tenths + 7 tenths = ____ tenths = ____

6 tenths + 6 tenths = ____ tenths = ____

5 tenths + 8 tenths = ____ tenths = ____

7 tenths + 4 tenths = ____ tenths = ____

3 tenths + 9 tenths = ____ tenths = ____

9 tenths + 8 tenths = ____ tenths = ____

ADDING AND SUBTRACTING TENTHS

Shade in the correct number of tenths and add.

□□□□□□□□□□ + □□□□□□□□□□ = □□□□□□□□□□ □□□□□□□□□□

0.8
+ 0.4

□□□□□□□□□□ + □□□□□□□□□□ = □□□□□□□□□□ □□□□□□□□□□

0.5

Add:

0.4 + 0.3	0.5 + 0.8	0.5 + 0.9	0.8 + 0.7	0.4 + 0.7
0.4 0.5 + 0.3	0.6 0.8 + 0.3	0.2 0.9 + 0.5	0.7 0.1 + 0.9	0.5 0.7 + 0.4
0.2 0.3 0.5 + 0.2	0.8 0.7 0.6 + 0.5	0.9 0.7 0.5 + 0.3	0.8 0.6 0.4 + 0.2	0.4 0.8 0.5 + 0.7

Subtract:

0.9 - 0.2	0.8 - 0.2	0.8 - 0.3	0.9 - 0.3	1.0 - 0.1
0.7 - 0.2	0.4 - 0.3	0.9 - 0.4	0.6 - 0.1	0.5 - 0.1

PICTURING HUNDREDTHS IN ADDITION AND SUBTRACTION

The example shows how we add hundredths with pictures.

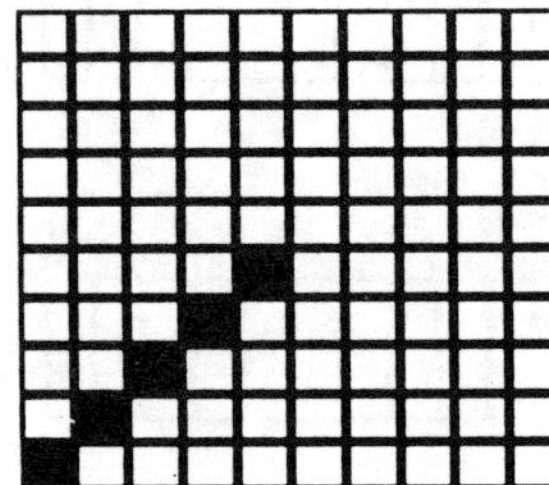 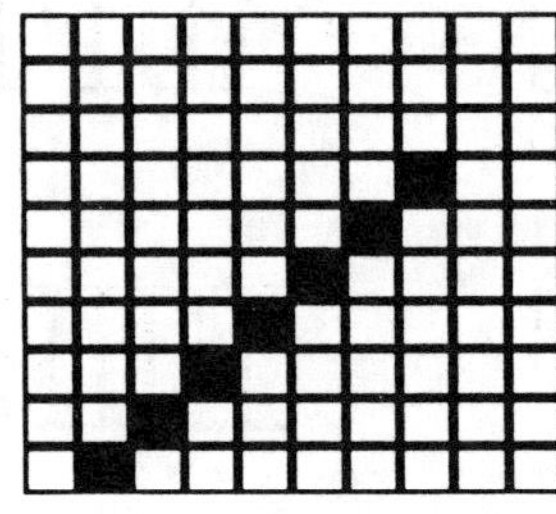 

5 hundredths + 7 hundredths = 12 hundredths = 0.12

We can add hundredths without pictures. Try these.

6 hundredths	+	9 hundredths	=	15	hundredths	=	0.15
12 hundredths	+	6 hundredths	=	____	hundredths	=	____
25 hundredths	+	5 hundredths	=	____	hundredths	=	____
43 hundredths	+	18 hundredths	=	____	hundredths	=	____

The example shows how we subtract hundredths with pictures.

 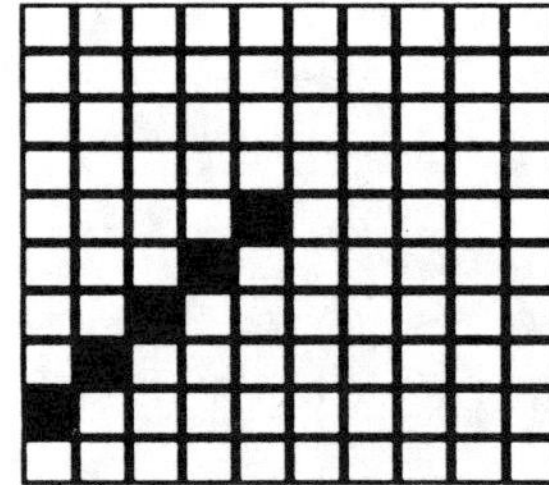 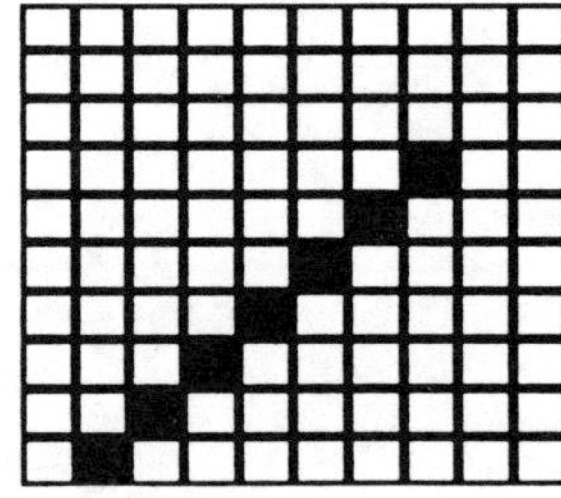

12 hundredths - 5 hundredths = 7 hundredths = ____

We can subtract hundredths without pictures. Try these.

36 hundredths	-	12 hundredths	=	24	hundredths	=	____
50 hundredths	-	25 hundredths	=	____	hundredths	=	____
72 hundredths	-	14 hundredths	=	____	hundredths	=	____
95 hundredths	-	63 hundredths	=	____	hundredths	=	____
46 hundredths	-	9 hundredths	=	____	hundredths	=	____

ADDING AND SUBTRACTING HUNDREDTHS

Shade in the correct number of hundredths and add.

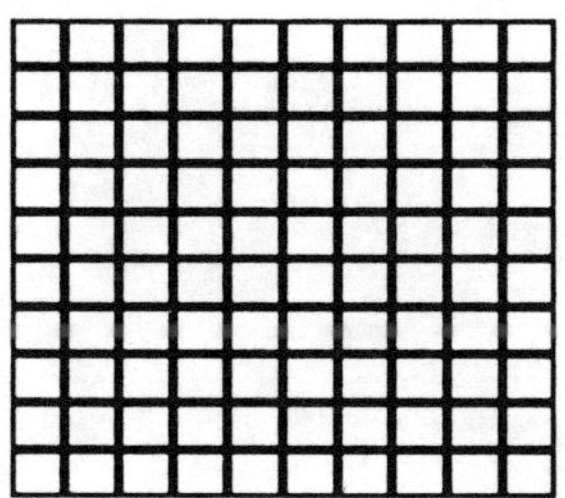 + =

0.21
+ 0.36

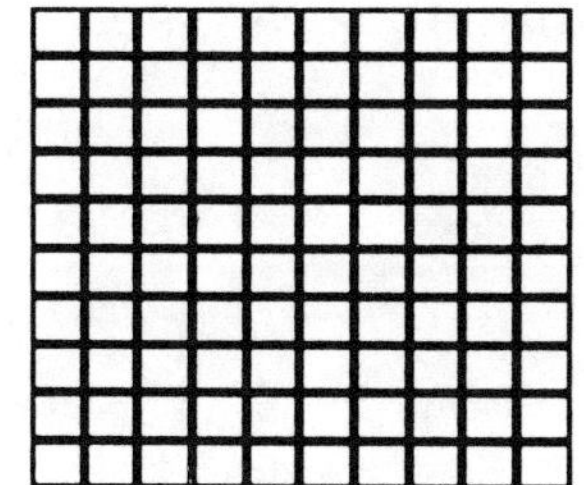 + = 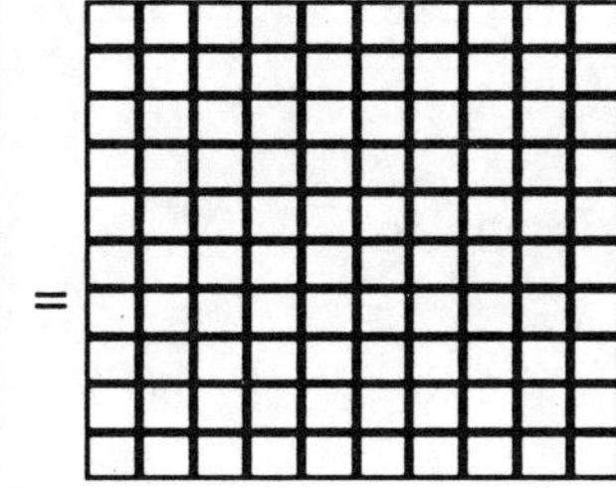 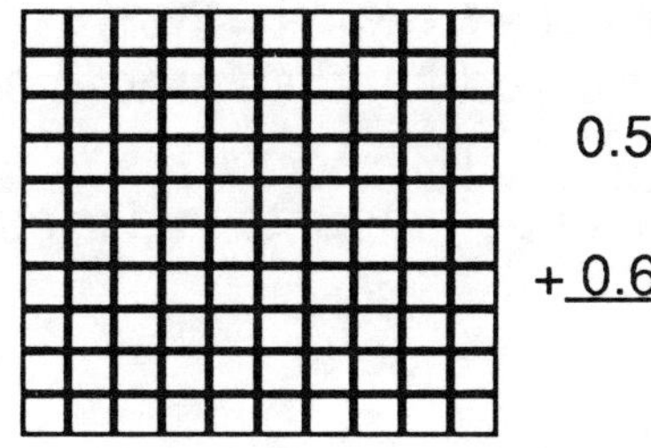

0.51
+ 0.62

Add or subtract as the sign indicates.

0.41	0.11	0.46	0.25	0.37
+ 0.52	+ 0.83	+ 0.39	+ 0.73	+ 0.46

0.81	0.92	0.73	0.66	0.23
- 0.42	- 0.16	- 0.44	- 0.57	- 0.07

Add:

0.46	0.57	0.64	0.71	0.83
+ 0.82	+ 0.63	+ 0.59	+ 0.72	+ 0.49

0.16	0.27	0.38	0.09	0.98
0.25	0.09	0.05	0.58	0.41
+ 0.83	+ 0.98	+ 0.86	+ 0.62	+ 0.17

MONEY IN ADDITION AND SUBTRACTION

When we add and subtract hundredths, we can think of adding and subtracting cents in the United States money system.

1 cent or **1¢** are other ways of writing one cent. **$0.01** is the decimal money form that we use when we add and subtract money.

Write these amounts in decimal money form:

3 cents	=	$0.03	24¢	=	$0.24	15 cents	=	$0.15
19 cents	=	$____	50¢	=	$____	42 cents	=	$____
81 cents	=	$____	8¢	=	$____	27 cents	=	$____
5¢	=	$____	99¢	=	$____	18¢	=	$____

Did you remember to place the decimal point in each amount above?

Add or subtract as the sign indicates. Write your answers in decimal money form.

6 cents	+	9 cents	=	15 cents	=	$0.15
12 cents	+	8 cents	=	______	=	$______
28 cents	-	5 cents	=	______	=	$______
92 cents	-	36 cents	=	______	=	$______
89 cents	-	23 cents	=	______	=	$______
21¢	+	16¢	=	______	=	$______
72¢	+	9¢	=	______	=	$______
65¢	-	24¢	=	______	=	$______
44¢	-	19¢	=	______	=	$______

ADDING AND SUBTRACTING AMOUNTS LESS THAN ONE DOLLAR

Write these amounts in decimal money form and add:

28 cents + 41 cents	= $0.28 + 0.41	23 cents + 8 cents =	$0.23 + 0.08
41 cents + 24 cents	= $ + ____	53 cents + 19 cents =	$ + ____
7¢ + 3¢ + 16¢	= $ + ____	34¢ + 5¢ + 9¢ =	$ + ____

Be sure to keep your decimal points in a straight line.

Add:

$ 0.54 + 0.37	$ 0.42 + 0.53	$ 0.62 + 0.29	$ 0.65 + 0.18	$ 0.41 + 0.29
$ 0.12 0.43 + 0.05	$ 0.09 0.52 + 0.18	$ 0.31 0.11 + 0.28	$ 0.19 0.53 + 0.17	$ 0.41 0.18 + 0.22

Subtract:

$ 0.84 - 0.15	$ 0.99 - 0.62	$ 0.51 - 0.25	$ 0.63 - 0.14	$ 0.85 - 0.07

The decimal point and the dollar sign should be in every answer.

ADDING AND SUBTRACTING DOLLARS AND CENTS

1 dollar and **$1** are other ways of writing one dollar. **$1.00** is the decimal money form that we use when we add and subtract money.

Write these amounts in decimal money form and add:

$3 + $4 + $6 = $ 3.00
4.00
+ 6.00

$13.00

$5 + $12 + 32¢ = $ 5.00
12.00
+ 0.32

$ ____

$7 + 33¢ + 56¢ = $

+____

$ ____

$50 + 82¢ + 24¢ = $

+____

$ ____

81¢ + 19¢ + 43¢ = $

+____

$ ____

$28 + 52¢ + 61¢ = $

+____

$ ____

Add these amounts:

$ 0.63	$ 3.21	$ 4.27	$ 6.27	$ 1.59
0.47	2.23	2.34	0.91	2.37
+ 0.52	+ 0.53	+ 8.41	+18.43	+ 11.84

$ 4.23	$ 2.59	$ 8.27	$ 80.42	$ 35.21
1.28	1.24	1.59	21.65	0.49
+ 2.84	+ 4.62	+ 2.43	+ 12.15	+ 33.56

Subtract these amounts:

$ 1.56	$ 22.59	$ 40.62	$ 52.16	$ 82.14
- 0.43	- 3.45	- 1.25	- 12.73	- 63.08

Take Quiz 3.

MONEY IN ACTION

A married couple used a checkbook for paying all of their bills. Add each deposit and subtract each check to find their ending balance.

CHECK NUMBER	DATE	DESCRIPTION	CHECK (-)	DEPOSIT (+)	BALANCE
	Jan. 1	Balance forward			$423.17
	1	Deposit - pay		1,559.27	
1058	1	Rent and taxes	645.00		
1059	3	Cash	100.00		
1060	4	Gas and Elec. Co.	57.81		
1061	5	Oil Company	38.62		
1062	7	Food Store	102.97		
	7	Deposit - pay		1,092.00	
1063	8	Clothing Store	72.49		
1064	9	Savings	200.00		
1065	11	Telephone Co.	38.45		
1066	12	Dr. Peters	30.00		
1067	13	Dental Clinic	40.00		
1068	14	Food Store	61.00		
1069	15	Cash Expenses	100.00		
1070	31	Food Store	200.00		
TOTAL CHECKS AND DEPOSITS					

MAKING THE BALANCE AGREE

Check book balance on January 1		$ 423.17
Add the total deposits	+	_______
Subtract the total checks written	-	_______
Checkbook balance on January 31		

MONEY PUZZLE

Solve the problems to unlock the puzzle.

A.	$5	+	$6.02	+	$5.24	=	________
E.	$24.23	+	$0.83	+	$3.19	=	________
H.	$6.72	+	$0.98	+	$1.59	=	________
I.	$80.76	-	$20.92			=	________
L.	$4.26	+	$9.27	+	$0.85	=	________
M.	$16.41	-	$2.58			=	________
N.	$0.56	+	$0.72	+	$6.19	=	________
O.	$164.01	-	$27.37			=	________
R.	$41.55	+	$76.22	+	$19.79	=	________
S.	$52	+	$63	+	$7.24	=	________
X.	$19.11	-	$2.57			=	________
Z.	$6.26	+	$1.83	+	$5.24	=	________

Look at your answers above and wrire the letters that represents each amount.

$16.26 $13.83-$59.84-$14.38-$14.38-$59.84-$136.64-$7.47

____ ____ ____ ____ ____ ____ ____ ____

$9.29-$16.26-$122.24 $122.24-$59.84-$16.54

____ ____ ____ ____ ____ ____

$13.33-$28.25-$137.56-$136.64-$122.34

____ ____ ____ ____ ____

KEEPING DECIMALS IN LINE IN ADDITION

When we add and subtract decimals, it is sometimes useful to add zeros to numbers so that we can keep the digits in a straight line.

We know that adding zeros to digits after the decimal point does not change the value of the decimal.

ALL OF THESE DECIMALS ARE EQUAL: -

0.2 = 0.20 = 0.200

Write these addition problems in the boxes below and add. Add zeros where necessary. An example has been done for you.

A.	1.203	+	0.52	+	0.9	**B.**	0.22	+	1.115	+	0.12
C.	6.009	+	1.2	+	4.18	**D.**	3.7	+	2.224	+	6.59
E.	12.3	+	1.66	+	2.121	**F.**	61.22	+	0.5	+	0.443
G.	21.35	+	6.9	+	2.03	**H.**	10.5	+	2.407	+	6.813

I. 41.3 + 2.07 + 3.149

A.
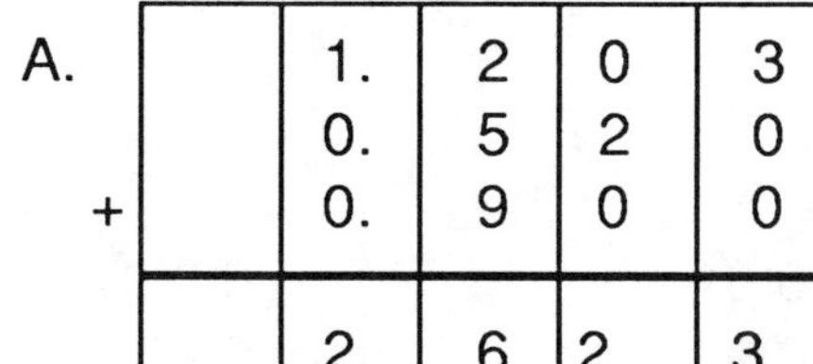

	1.	2	0	3
	0.	5	2	0
+	0.	9	0	0
	2.	6	2	3

B.

C.
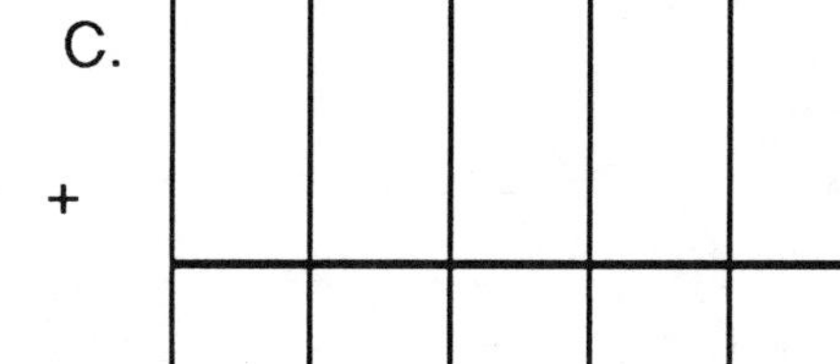

D.

E.
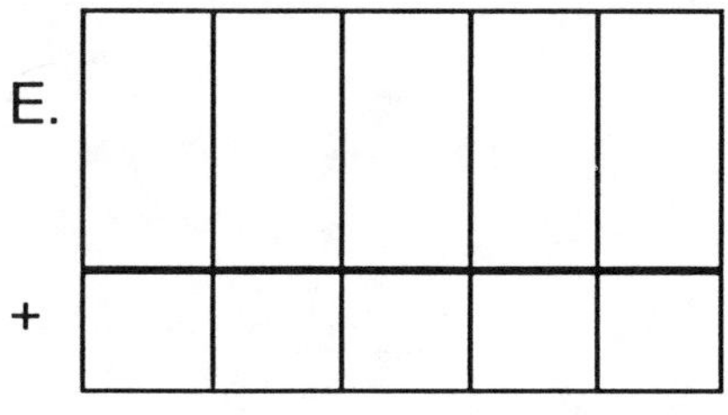

F.
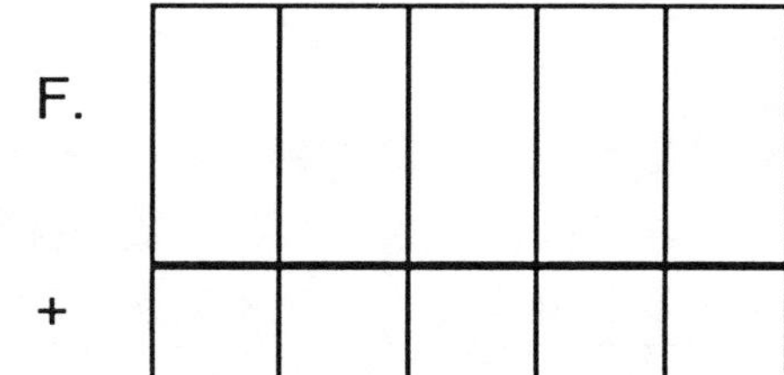

G.
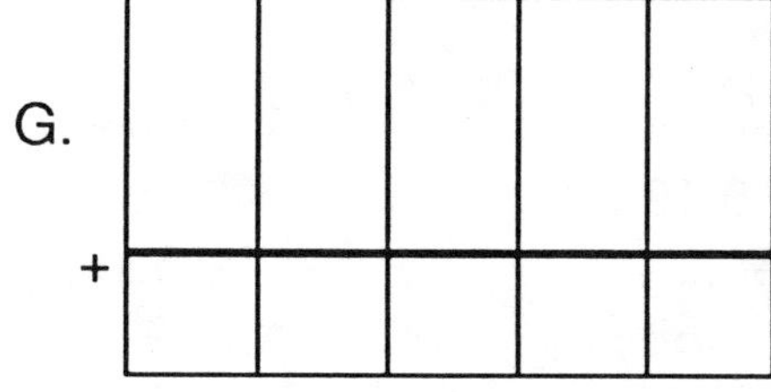

H.
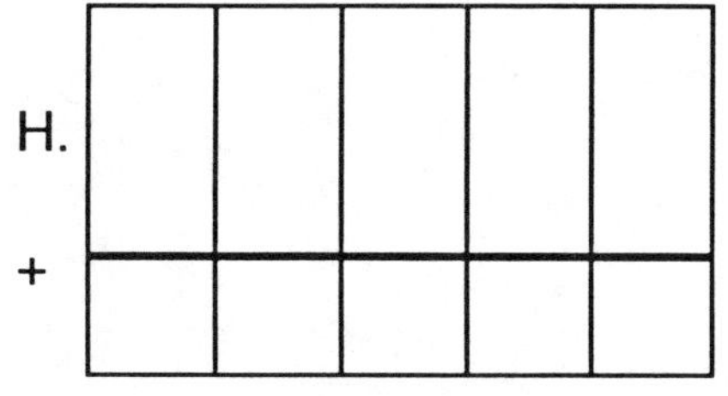

I.
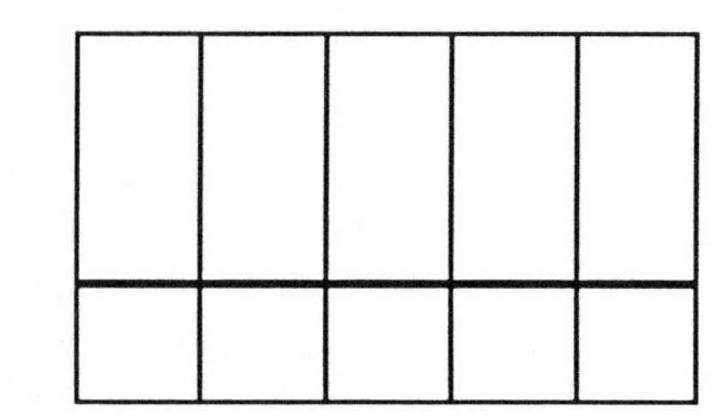

PRACTICE PAGE IN ADDITION

Write these decimals in a straight line and add. Add zeros where necessary.

0.1 + 0.03 + 0.104 =
```
  0.100
  0.030
+ 0.104
  0.234
```

0.5 + 0.07 + 0.222 =
```
  0.500
  0.070
+ 0.222
```

1.21 + 0.507 + 1.5 =

```
+ _____
  _____
```

2.7 + 1.83 + 0.407 =

```
+ _____
  _____
```

3.4 + 9.18 + 2.653 =

```
+ _____
  _____
```

4.69 + 1.103 + 0.6 =

```
+ _____
  _____
```

Add:

```
   0.01        0.09        0.6        0.81        0.003
 + 0.5       + 0.503     + 0.72     + 0.9       + 0.62
 ------      -------     ------     ------      ------

   1.2         2.81        2.5        6.31        1.32
   3.72        4.051       7.6        4.6         8.141
 + 8.914     + 5.1       + 8.16     + 2.006     + 6.8
 -------     ------      ------     -------     ------

   8.1         7.33        9.207      5.193       2.45
   2.66        1.9         0.3       64.9        11.52
 + 5.009     + 2.175      +3.22     + 4.888       1.006
 -------     -------     ------     -------     ------
```

KEEPING DECIMALS IN LINE IN SUBTRACTION

In subtraction, it is often necessary to regroup. Adding zeros to digits after the decimal point makes this task much easier.

Write these subtraction problems in the boxes below and subtract. Add zeros where necessary. An example has been done for you.

A.	0.4	-	0.21	B.	0.43	-	0.123
C.	0.52	-	0.106	D.	2.3	-	1.12
E.	6.35	-	2.413	F.	8.92	-	2.115
G.	6.3	-	2.104	H.	17.5	-	3.426
I.	12.6	-	2.417	J.	6.8	-	1.207
K.	25	-	12.203	L.	40	-	10.789

A.
-
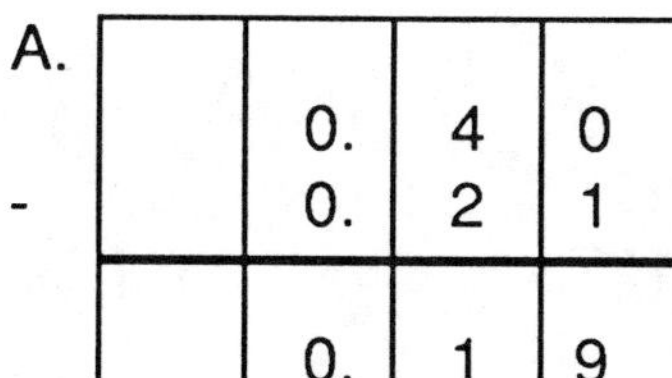

B.
-
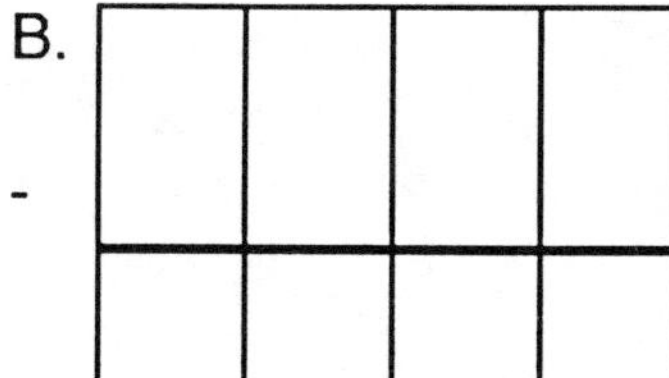

C.
-
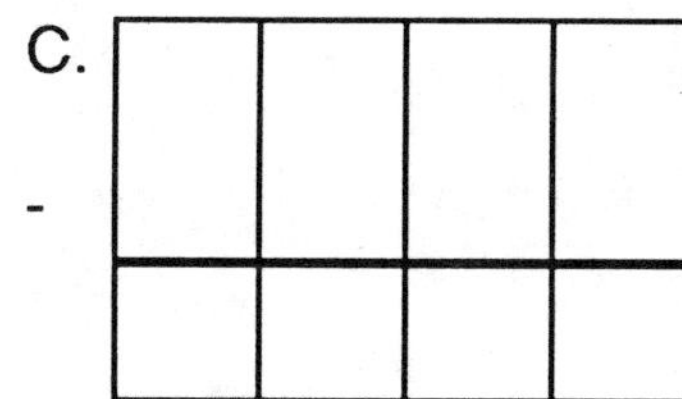

D.
-
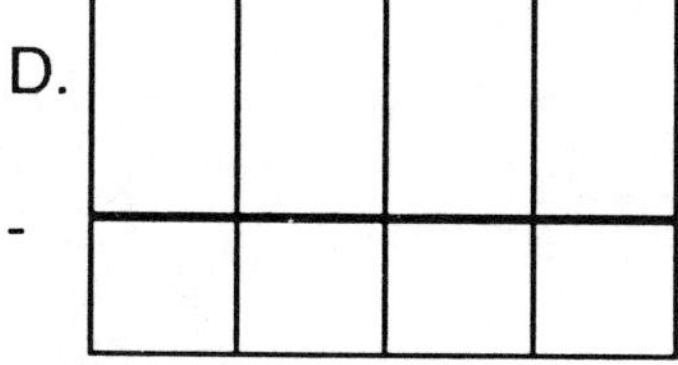

E.
-
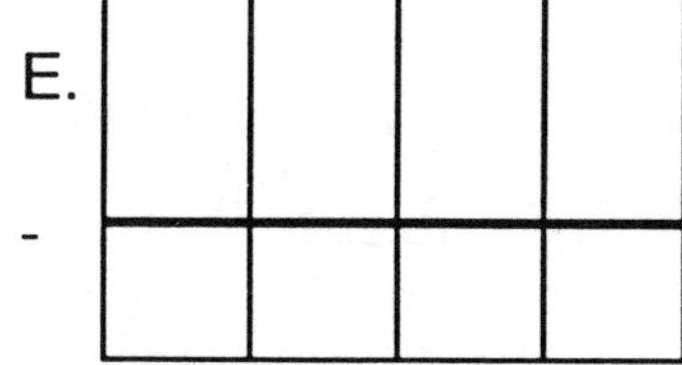

F.
-
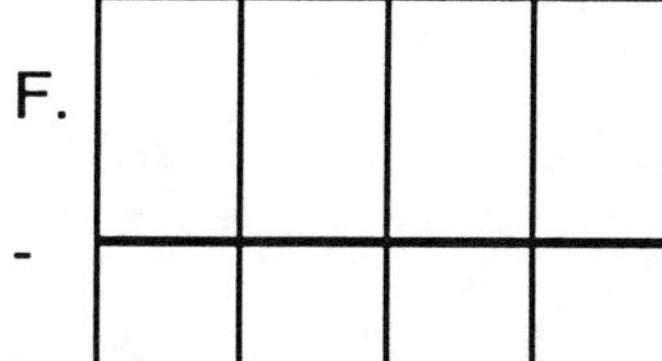

G.
-
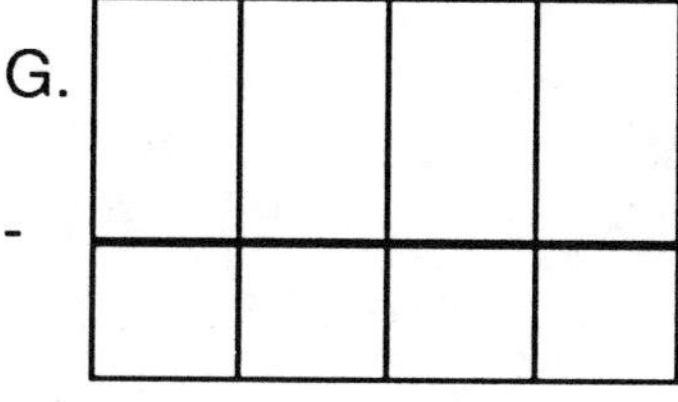

H.
-
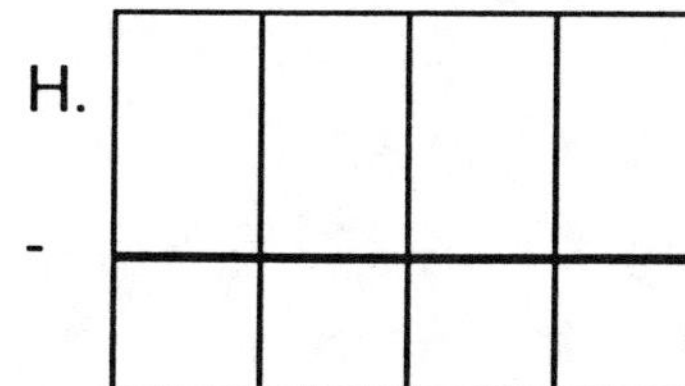

I.
-
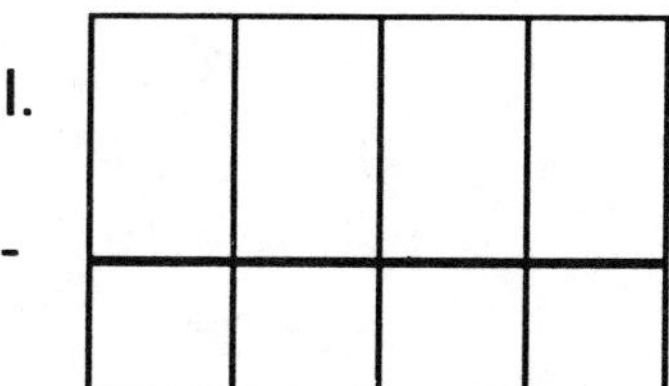

J.
-
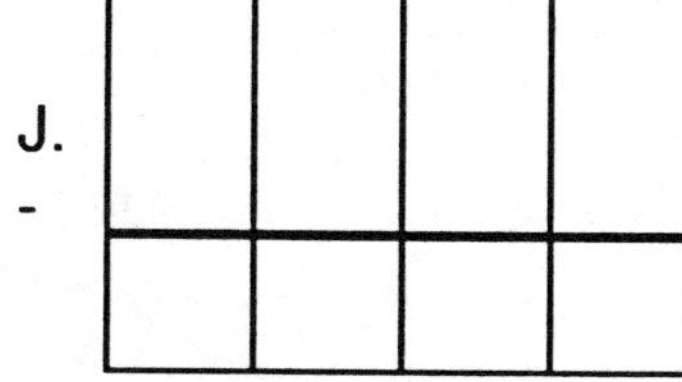

K.
-
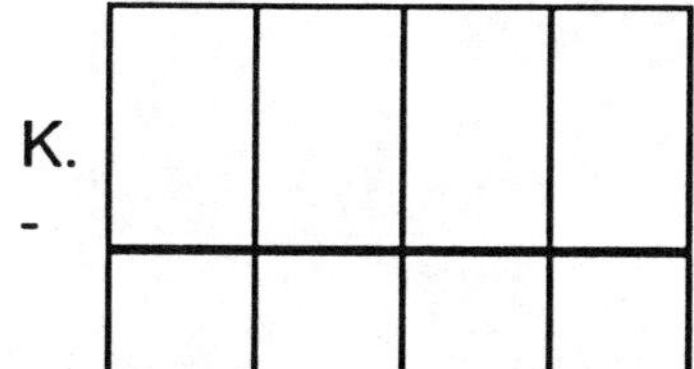

L.
-

PRACTICE PAGE IN SUBTRACTION

Write these decimals in a straight line and subtract. Add zeros where necessary.

0.63 - 0.271 = 0.630
- 0.271

0.359

0.53 - 0.124 = 0.530
- 0.124

7.23 - 1.409 =

18.6 - 1.207 =

18.3 - 11.207 =

20 - 1.5 =

Are your decimal points lined up?

Subtract:

0.23 - 0.105	2.8 - 1.23	16.3 - 2.11	24.23 - 12.127	60. - 1.3
10.407 - 1.9	76.3 - 23.217	84.92 - 3.158	72.4 - 1.276	51.8 - 2.691
4.8 - 1.302	27.6 - 16.318	14.52 - 2.483	32.65 - 1.427	65.71 - 8.246
6.3 - 1.444	28.5 - 14.865	99.1 - 52.678	82.3 - 7.865	44.3 - 5.663

NAME:______________________________

QUIZ 1

Write the fraction name and the decimal name for each picture.

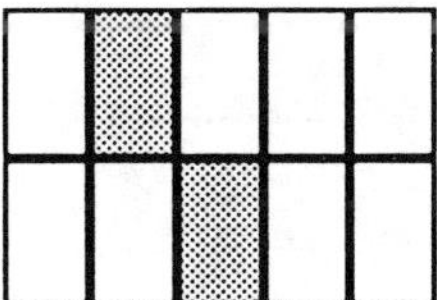

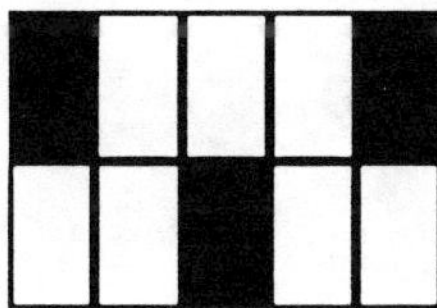

Fraction name: ________ ________ ________

Decimal name: ________ ________ ________

Write these whole numbers as decimal numbers:

3	=	_____	602	=	_____	82	=	_____
12	=	_____	75	=	_____	50	=	_____

Write these fractions as decimals:

6/10	=	_____	7/10	=	_____	4/10	=	_____
3/10	=	_____	1/10	=	_____	5/10	=	_____

Write these whole and mixed numbers as decimals:

1 6/10	=	_____	9 2/10	=	_____	8 9/10	=	_____
3 5/10	=	_____	7	=	_____	9	=	_____

Write these mixed numbers as fractions:

2.1	=	_____	6.9	=	_____	8.4	=	_____
6.6	=	_____	20.5	=	_____	9.3	=	_____

NAME:______________________________

QUIZ 2

Write these fractions as decimals.

3/100 = ________	43/100 = ________	79/100 = ________
4/1000 = ________	27/1000 = ________	727/1000 = ________

Write these mixed numbers as decimals.

1 4/100 = ________	13 71/100 = ________	12 17/100 = ________
6 2/1000 = ________	50 363/1000 = ________	14 21/1000 = ________

Write these decimals as fractions.

0.04 = ________	0.19 = ________	0.28 = ________
0.121 = ________	0.23 = ________	0.001 = ________
0.56 = ________	0.179 = ________	0.2 = ________

Write these mixed numbers as fractions.

21.04 = ________	2.89 = ________	1.05 = ________
1.002 = ________	12.123 = ________	98.068 = ________

Compare the first decimal with the second decimal. Use **<** or **>**.

0.2 ______ 0.5	0.61 ______ 0.05	0.213 ______ 0.021
0.7 ______ 0.04	0.82 ______ 0.9	0.51 ______ 0.83
0.123 ______ 0.5	0.269 ______ 0.32	0.66 ______ 0.7

NAME:______________________

QUIZ 3

Add or subtract as the sign indicates.

0.7 + 0.4	0.8 - 0.2	0.9 + 0.8	0.9 - 0.3	0.8 + 0.7
0.8 0.5 0.6 + 0.3	0.7 0.2 0.5 + 0.3	0.8 0.7 0.6 + 0.5	0.2 0.4 0.2 + 0.6	0.4 0.6 0.8 + 0.3
0.45 + 0.52	0.86 - 0.23	0.53 + 0.22	0.95 - 0.23	0.63 + 0.24
0.61 0.34 + 0.52	0.64 0.57 + 0.38	0.09 0.83 + 0.75	0.26 0.05 + 0.99	0.65 0.27 + 0.48
$ 0.47 + 0.29	$ 0.82 - 0.46	$ 0.75 + 0.16	$ 0.84 - 0.47	$ 0.45 + 0.48
$ 15.42 3.45 + 0.65	$ 1.27 3.28 + 4.19	$ 20.65 2.53 + 3.18	$ 6.37 11.23 + 1.48	$ 16.52 1.82 + 9.04

Be sure that you have placed a decimal point in all of your answers. Did you include a dollar sign in the money problems? _____

NAME:______________________________

QUIZ 4

Add or subtract as the sign indicates.

1.206 + 3.2 + 4.51 = ____________________

22.3 + 4.62 + 0.392 = ____________________

2.005 + 5.1 + 6.09 = ____________________

15 + 0.51 + 6.1 + 0.003 = ____________________

0.521 + 11 + 16.3 + 42.91 = ____________________

62.1 - 3.49 = ____________________

18.23 - 1.007 = ____________________

19.9 - 4.231 = ____________________

15 - 0.038 = ____________________

77 - 2.51 = ____________________

6.5	3.206	62.6	35.34	9.26
1.92	14.3	1.009	6.1	1.045
+ 3.106	+ 18.92	+ 8.32	+ 8.005	+ 3.18

82.6	14.62	24.3	16.	23.
- 1.31	- 1.096	- 1.271	- 1.204	- 1.09

SELECTED ANSWERS

PAGE 4

8 tenths
4 hundredths
9 thousandths

PAGE 9

0.5	0.9	0.8
0.7	0.4	0.1
0.3	0.6	0.2

17.2	18.7	33.8
11.0	29.4	128.1
25.3	2.5	93.0

2 1/10	19 6/10	8/10
15 1/10	7/10	123 6/10
17 4/10	3/10	12 8/10

point, point, decimal, tenths

PAGE 14

0.04	0.08	0.45
0.19	0.34	0.15
0.03	0.07	0.11
0.66	0.53	0.91

4/100	75/100	82/100
2/100	55/100	64/100
12/100	9/100	13/100

PAGE 17

	0.016	0.032
0.004	0.122	0.065
0.081	0.005	0.044
0.002	0.075	0.343

3/1000	12/1000	15/1000
875/1000	128/1000	64/1000

PAGE 17

9/1000	102/1000	93/1000
98/1000	644/1000	333/1000

PAGE 19

THIS IS A PUZZLE!

PAGE 31

0.7	1.3	1.4	1.5	1.1
1.2	1.7	1.6	1.7	1.6
1.2	2.6	2.4	2.0	2.4

0.7	0.6	0.5	0.6	0.9
0.5	0.1	0.5	0.5	0.4

PAGE 33

0.93	0.94	0.85	0.98	0.83
0.39	0.76	0.29	0.09	0.16
1.28	1.20	1.23	1.43	1.32
1.24	1.34	1.29	1.29	1.56

PAGE 35

		$0.69	$0.31	
		$0.65	$0.72	
		$0.26	$0.48	
$0.91	$0.95	$0.91	$0.83	$0.70
$0.60	$0.79	$0.70	$0.89	$0.81
$0.69	$0.37	$0.26	$0.49	$0.78

PAGE 36

			$17.32	
		$7.89	$51.06	
		$1.43	$29.13	
$1.62	$5.97	$15.02	$25.61	$15.80
$8.35	$8.45	$12.29	$114.22	$69.26
$1.13	$19.14	$39.37	$39.43	$19.06

PAGE 37

Running balances:

		$423.17	$1982.44
$1337.44	$1237.44	$1179.63	$1141.01
$1038.04	$2130.04	$2057.55	$1857.55
$1819.10	$1789.10	$1749.10	$1688.16
$1588.10	$1388.10		

Making the Balance Agree:

Check book balance on January 1		$ 423.17
Deposits	+	2651.27
		$ 3074.44
Checks written	-	1686.34
Balance on January 31		$ 1388.10

SELECTED ANSWERS

PAGE 38

$16.26	$28.25	$9.29
$59.84	$14.38	$13.83
$7.47	$136.64	$137.56
$122.24	$16.54	$13.33

A MILLION HAS SIX ZEROS

PAGE 40

		3.217	4.937	
		15.233	6.393	
0.51	0.593	1.32	1.71	0.623
13.834	11.961	18.26	12.916	16.261
15.769	11.405	12.727	74.981	14.976

PAGE 42

			0.406	
		5.821	17.393	
		7.093	18.5	
0.125	1.57	14.19	12.103	58.7
8.507	53.083	81.762	71.124	49.109
3.498	11.282	12.037	31.223	57.464
4.856	13.635	46.422	74.435	38.637

QUIZ 3

1.1	0.6	1.7	0.6	1.5
2.2	1.7	2.6	1.4	2.1
0.97	0.63	0.75	0.72	0.87
1.47	1.59	1.67	1.30	1.40
$0.76	$0.36	$0.91	$0.37	$0.93
$19.52	$8.74	$26.36	$19.08	$27.38

QUIZ 4

		8.916		
		27.312		
		13.195		
		21.613		
		70.731		
		58.61		
		17.223		
		15.669		
		14.962		
		74.49		
11.526	36.426	71.929	49.445	13.485
81.29	13.524	23.029	14.796	21.91

PAGE 39

	1.455	11.389
12.514	16.081	62.163
30.28	19.720	46.519

PAGE 41

	0.307	0.414
1.18	3.937	6.805
4.196	14.074	10.183
5.593	12.797	29.211

QUIZ 1

2/10	5/10	3/10
0.2	0.5	0.3
3.0	602.0	82.0
12.0	75.0	50.0
0.6	0.7	0.4
0.3	0.1	0.5
1.6	9.2	8.9
3.5	7.0	9.0
2 1/10	6 9/10	8 4/10
6 6/10	20 5/10	9 3/10

QUIZ 2

0.03	0.43	0.79
0.004	0.027	0.727
1.04	13.71	12.17
6.002	50.363	14.021
4/100	19/100	28/100
121/1000	23/100	1/1000
56/100	179/1000	2/10
21 4/100	2 89/100	1 5/100
1 2/1000	12 123/1000	98 68/1000
<	>	>
>	<	<
<	<	<